AF749322

Antiphospholipid Syndrome - Recent Advances in Clinical and Basic Aspects

Edited by Polona Žigon

Published in London, United Kingdom

IntechOpen

Supporting open minds since 2005

Antiphospholipid Syndrome - Recent Advances in Clinical and Basic Aspects
http://dx.doi.org/10.5772/intechopen.91508
Edited by Polona Žigon

Contributors
Pavol Holly, Peter Kubisz, Jan Stasko, Polona Žigon, Ula Štok, Snežna Sodin-Šemrl, Saša Čučnik, Kenji Tanimura, Hideto Yamada, Yuki Sasagawa, Masashi Deguchi, Noriko Arase, Hisashi Arase, John Cole, Matthew J. Woodward, Sana Somani, Adeolu Morawo, Rakhee Lalla, Adam P. Klein, Kathryn Grimes, Ariela Hoxha, Paolo Simioni

© The Editor(s) and the Author(s) 2022
The rights of the editor(s) and the author(s) have been asserted in accordance with the Copyright, Designs and Patents Act 1988. All rights to the book as a whole are reserved by INTECHOPEN LIMITED. The book as a whole (compilation) cannot be reproduced, distributed or used for commercial or non-commercial purposes without INTECHOPEN LIMITED's written permission. Enquiries concerning the use of the book should be directed to INTECHOPEN LIMITED rights and permissions department (permissions@intechopen.com).
Violations are liable to prosecution under the governing Copyright Law.

Individual chapters of this publication are distributed under the terms of the Creative Commons Attribution 3.0 Unported License which permits commercial use, distribution and reproduction of the individual chapters, provided the original author(s) and source publication are appropriately acknowledged. If so indicated, certain images may not be included under the Creative Commons license. In such cases users will need to obtain permission from the license holder to reproduce the material. More details and guidelines concerning content reuse and adaptation can be found at http://www.intechopen.com/copyright-policy.html.

Notice
Statements and opinions expressed in the chapters are these of the individual contributors and not necessarily those of the editors or publisher. No responsibility is accepted for the accuracy of information contained in the published chapters. The publisher assumes no responsibility for any damage or injury to persons or property arising out of the use of any materials, instructions, methods or ideas contained in the book.

First published in London, United Kingdom, 2022 by IntechOpen
IntechOpen is the global imprint of INTECHOPEN LIMITED, registered in England and Wales, registration number: 11086078, 5 Princes Gate Court, London, SW7 2QJ, United Kingdom
Printed in Croatia

British Library Cataloguing-in-Publication Data
A catalogue record for this book is available from the British Library

Additional hard and PDF copies can be obtained from orders@intechopen.com

Antiphospholipid Syndrome - Recent Advances in Clinical and Basic Aspects
Edited by Polona Žigon
p. cm.
Print ISBN 978-1-83969-053-2
Online ISBN 978-1-83969-054-9
eBook (PDF) ISBN 978-1-83969-055-6

For EU product safety concerns:
IN TECH d.o.o., Prolaz Marije Krucifikse Kozulić 3, 51000 Rijeka, Croatia, info@intechopen.com or visit our website at intechopen.com.

We are IntechOpen, the world's leading publisher of Open Access books

Built by scientists, for scientists

5,800+
Open access books available

142,000+
International authors and editors

180M+
Downloads

156
Countries delivered to

Our authors are among the
Top 1%
most cited scientists

12.2%
Contributors from top 500 universities

CLARIVATE ANALYTICS
BOOK CITATION INDEX
INDEXED

WEB OF SCIENCE™

Selection of our books indexed in the Book Citation Index (BKCI)
in Web of Science Core Collection™

Interested in publishing with us?
Contact book.department@intechopen.com

Numbers displayed above are based on latest data collected.
For more information visit www.intechopen.com

Meet the editor

Dr. Polona Žigon is a senior scientist at the Department of Rheumatology, University Medical Center Ljubljana, Slovenia, and is a teaching assistant at the Faculty of Mathematics, Natural Sciences, and Information Technologies, University of Primorska, Slovenia. She obtained an MSc in Clinical Biochemistry in 2008 and a Ph.D. in Biomedicine in 2014 at the Faculty of Pharmacy, University of Ljubljana. During her post-graduate studies, she completed a six-month research internship in the School of Medicine, Hokkaido University Sapporo, Japan. She is a member of the Slovene Chamber Laboratory Medicine, a working member of EMEUNET, and a member of the Autoimmunity Network. Her research in the field of systemic autoimmune diseases resulted in more than thirty original scientific articles published in peer-reviewed journals, six book chapters, and one patent application.

Contents

Preface XI

Section 1
Introduction 1

Chapter 1 3
Introductory Chapter: Antiphospholipid Antibodies - A Laboratory Criterion for the Antiphospholipid Syndrome, but Also Bystanders in Infections, Cancer, and Other Conditions
by Polona Žigon

Section 2
Clinical Aspects of the Antiphospholipid Syndrome 17

Chapter 2 19
Antiphospholipid Syndrome and Stroke
by Kathryn Grimes, Adam P. Klein, Rakhee Lalla, Adeolu Morawo, Sana Somani, Matthew J. Woodward and John W. Cole

Chapter 3 49
Obstetric Antiphospholipid Syndrome
by Ariela Hoxha and Paolo Simioni

Chapter 4 71
Bleeding in Patients with Antiphospholipid Antibodies
by Peter Kubisz, Pavol Holly and Jan Stasko

Section 3
Trends and Novelties in the Diagnosis and Pathogenesis of Antiphospholipid Syndrome 91

Chapter 5 93
A Novel Autoantibody against β2-Glycoprotein I/HLA Class II Complexes in Antiphospholipid Syndrome
by Kenji Tanimura, Yuki Sasagawa, Masashi Deguchi, Noriko Arase, Hisashi Arase and Hideto Yamada

Chapter 6 103
Extracellular Vesicles: Intercellular Communication Mediators in Antiphospholipid Syndrome
by Ula Štok, Saša Čučnik, Snežna Sodin-Šemrl and Polona Žigon

Preface

Antiphospholipid syndrome (APS) is a systemic autoimmune disorder characterized by vascular thrombosis and/or pregnancy-related morbidity in the presence of persistently positive antiphospholipid antibodies (aPL). APS is considered the most common acquired form of thrombophilia worldwide. Obstetric APS is a complex entity that can affect both the mother and the fetus throughout pregnancy with high morbidity. At least one clinical criterion (vascular thrombosis or pregnancy morbidity) and one laboratory-based criterion (positive test result for lupus anticoagulant, anticardiolipin antibodies, and/or anti-β2-glycoprotein-1 antibodies) must be met for a patient to be classified as having APS. In the general population, the incidence of clinical manifestations significant for APS is high and could often be triggered by other underlying factors. Therefore, the diagnosis of APS relies primarily on laboratory measurements of aPL. However, current laboratory-based tests for aPL are hampered by technical limitations. Despite numerous attempts to increase their specificity, a high number of patients are still misdiagnosed. There is a need for novel, robust, and reliable biomarkers to firstly detect APS and secondly to monitor the risk for recurrent events over time.

This book consists of three sections. The first section contains the introductory chapter. The second section discusses important clinical aspects of APS and the cellular and/or molecular mechanisms potentially involved. Topics covered in this section include stroke and APS, obstetric manifestations of APS, and bleeding complications in APS. The third section discusses novelties in the diagnosis and pathogenesis of APS. The two chapters in this section examine the diagnostic utility of a novel autoantibody against β2-glycoprotein I/HLA class II complexes and recent findings in the field of extracellular vesicles, which offer promising aspects that may explain their role in the pathogenesis of APS.

Polona Žigon
Department of Rheumatology,
University Medical Centre Ljubljana,
Ljubljana, Slovenia

Section 1

Introduction

Chapter 1

Introductory Chapter: Antiphospholipid Antibodies - A Laboratory Criterion for the Antiphospholipid Syndrome, but Also Bystanders in Infections, Cancer, and Other Conditions

Polona Žigon

1. Introduction

Antiphospholipid syndrome (APS) is a systemic autoimmune thromboinflammatory disorder characterized by vascular thrombosis and pregnancy-related morbidity accompanied by persistent positive antiphospholipid antibodies (aPL) [1, 2]. APS is considered the most common acquired form of thrombophilia worldwide [3]. Obstetric APS is a complex entity that can affect both the mother and the fetus throughout pregnancy with high morbidity. The clinical complications of obstetric APS are diverse and include recurrent fetal loss, stillbirth, intrauterine growth failure, and preeclampsia [4]. In addition to thrombosis and pregnancy loss, other pathological manifestations regularly occur with APS including thrombocytopenia, destruction of heart valves, accelerated atherosclerosis, nephropathy, movement disorders, and cognitive decline [5]. Catastrophic APS (CAPS) is characterized by the rapid development of thrombosis in multiple organs and micro-thrombosis within a short period of time. Pediatric APS is a rare condition that is distinctly different from adult APS [6].

The classification of APS for clinical trials and studies is currently based on the international consensus statement established in Sapporo in 1999 and updated in Sydney in 2006, and includes a clinical criterion (vascular thrombosis or pregnancy morbidity) and a laboratory criterion (positive test result for aPL) [1] as shown in **Figure 1**. aPL are a heterogeneous family of IgG and/or IgM or, more rarely, IgA autoantibodies with an affinity for negatively charged phospholipids or protein-phospholipid complexes. Their persistent presence in sera has been associated with increased prothrombotic risk in various autoimmune diseases. The aPL that constitute the laboratory criteria for APS include lupus anticoagulant (LA), anti-cardiolipin antibodies (aCL), and anti-β2-glycoprotein I antibodies (anti-β2GPI) of immunoglobulin IgG and IgM classes. Extensive evidence has accumulated over the past decade that several other than those included in the APS classification criteria may be relevant to APS pathogenesis. Among them, antiprothrombin antibodies, especially antibodies against phosphatidylserine-prothrombin complex (aPS/PT), are supported by the most studies in the literature showing their strong correlation to LA activity and to clinical manifestations of APS [7–9]. An international multidisciplinary initiative "APS action", jointly supported by the American College of

IntechOpen

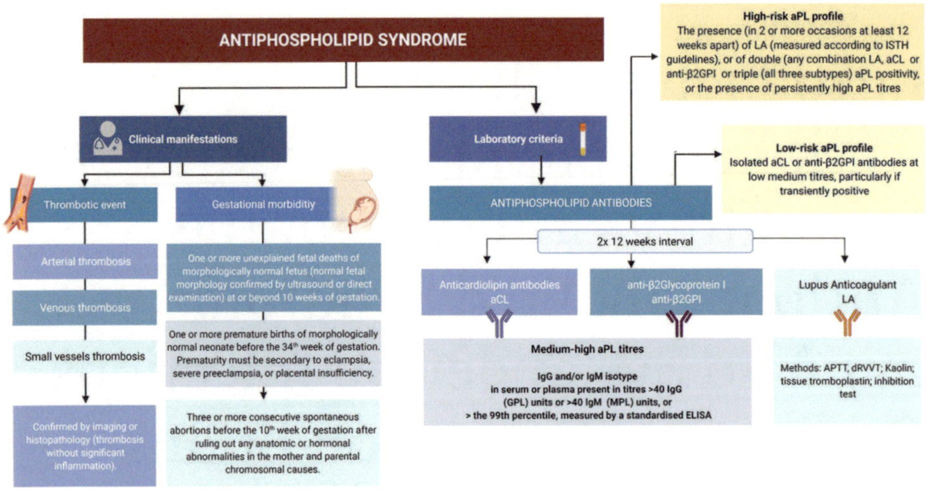

Figure 1.
Classification criteria for APS and definition of high and low risk profile. Created with BioRender.com.

Rheumatology (ACR) and the European League Against Rheumatism (EULAR) is currently underway to establish a new diagnostic criterion for APS.

APS can either be a disease in the absence of evidence of other autoimmune disease, or it can be secondary to another autoimmune disease such as systemic lupus erythematous (SLE) [10]. The profile of aPL, including type and titer, is an important factor determining the risk for thrombotic and obstetric events [11, 12]. The presence of LA, triple positivity or double positivity with positive LA, and the persistent presence of high titers of aCL and anti-β2GPI antibodies pose a high risk for the development of APS. In contrast, isolated positivity at low or medium titers of aCL or anti- β2GPI antibodies, particularly when transiently positive, poses a low risk.

A very rare, but life-threatening form of multiorgan thrombosis is known as catastrophic anti-phospholipid syndrome (CAPS) [13–15]. It is characterized by simultaneous thrombosis in multiple organs within a short period of time, that is, within a few days. Thrombosis often occurs at unusual sites, and small and medium-sized arteries are most frequently involved [16]. Less than 1% of patients with APS develop CAPS. CAPS is the first manifestation of APS in about half of diagnosed CAPS patients. The remaining patients have a history of APS. The mortality rate has decreased over time, mainly due to triple therapy (anticoagulation, corticotherapy and therapeutic plasma exchange—TPE—or intravenous immunoglobulin—IVIG), but it still exceeds 30% [17]. An international registry established in 2000 by the European Forum on Anti-Phospholipid Anti-bodies, and the last reported data (2016) includes 500 patients [17].

The other major clinical manifestations of APS are obstetric. These include unexplained death of one or more morphologically normal fetuses at or after 10-week gestation, premature delivery of one or more morphologically normal newborns before 34-week gestation due to either eclampsia, severe preeclampsia, or recognized features of placental insufficiency, and three or more unexplained, consecutive spontaneous abortions before 10-week gestation.

2. Clinical manifestations of APS are heterogeneous and nonspecific

The heterogeneity and non-specificity of potential clinical signs illustrates that APS is as a true systemic autoimmune disease and underscores the need for a better understanding of disease mechanisms that will enable a personalized approach to treatment. Despite some improvements in the diagnosis and prognosis of APS and the prevention of thrombosis reoccurrence, robust laboratory biomarkers are still lacking.

Because APS affects young patients in the most productive years of their lives, the consequences of organ or tissue damage lead to impaired health-related quality of life (HRQoL). There are several reasons why APS could adversely affect HRQoL. The clinical manifestations are diverse, and many of them damage vital tissues. In addition, APS may overlap with rheumatoid arthritis (RA) and SLE, both of which already significantly affect HRQoL. Another aspect that affects HRQoL in APS is treatment with high-dose anticoagulation indefinitely in patients with thrombosis and/or at high risk of thrombosis.

In the general population, the incidence of clinical manifestations present in APS is high and could often be triggered by other underlying factors. Consequently, the diagnosis of APS relies predominantly on laboratory measurements. However, current laboratory tests are hampered by technical limitations in the pre-analytical and analytical phases and by the fact that there is no standardization of these tests. Despite the many attempts to increase the specificity of laboratory criteria and the establishment of consensus criteria for serology, a high number of patients are still misdiagnosed. One of the most important reasons for this is the high heterogeneity of aPL in patients with APS. Thus, it remains to be clarified whether different manifestations are caused by subpopulations of autoantibodies against different epitope specificities that are currently detected by the same test(s). Unfortunately, most APS patients exhibit more than one type of aPL, making it difficult to assign pathogenic effects to one epitope specificity or another. In addition, the diagnosis of pediatric APS is even more challenging since it is such a rare condition. Diagnosis may be delayed or missed when adult APS criteria are used, because in pediatric APS, non-thrombotic clinical manifestations, such as thrombocytopenia, hemolytic anemia, and neurologic disorders such as migraine, epilepsy, and chorea, may precede thrombotic manifestations.

3. Current laboratory criteria are unable to identify all patients with APS

While aPL circulate in relatively stable concentrations in the blood, thrombosis occurs only occasionally. The persistent presence of aPL is thought to shift the hemostatic balance toward a prothrombotic state, but then a "second hit" is required to trigger the thrombotic event itself. Although this two-hit model is generally accepted, much remains to be learned about how exactly aPL predispose to thrombosis *in vivo* and how this predisposition interacts with the second hit. The pathogenic mechanisms responsible for thrombosis and obstetric complications in APS are based on a combination of factors, including inhibition of natural anticoagulant pathways, disruption of the annexin A5 anticoagulant shield on the phospholipid surface, activation of cellular elements, hemostatic reactions, and inflammation, particularly complement activation. Binding of aPL to the surface of vascular cells (endothelial cells, platelets, monocytes, trophoblasts) triggers activation of these cells, resulting in an increase in surface expression, production, and activity of procoagulant molecules and triggering the release of extracellular vesicles (EVs). EVs are submicron particles that are constitutively released by almost all cell type. In response to stimuli, such as cell activation by inflammation and/or apoptosis, their release into the cell surroundings is triggered to an even greater extend. EVs carry a diverse cargo (bioactive lipids, proteins, and nucleic acids) and might reflect the cell of origin and even its activation status. An increase in circulating EVs, particularly endothelial EVs, is considered a hallmark of vascular dysfunction and cardiovascular disease. Moreover, increased levels of EVs in the absence of an acute thrombotic event suggest a chronic state of vascular activation in APS. EVs could therefore be a useful biomarker to identify patients with aPL at the highest risk for complications. The lack of standardized approaches to isolate and/or characterize EVs has been a major limitation in determining their role in various diseases, including APS. Few studies have investigated EVs in APS patients. These studies have been limited to characterization of medium- to large-sized EVs, with significantly higher concentrations of endothelial and platelet EVs detected in the plasma of APS patients compared with healthy controls [9]. Despite well-characterized in vitro models of APS pathology, the field of EVs remains largely unexplored and may therefore provide insight into the APS mechanism. To our knowledge, no study has investigated whether EVs isolated from the vicinity of aPL-stimulated cells have the potential to activate distant endothelial cells in a similar manner.

In the general population, the incidence of clinical manifestations which can be attributable to APS is high and could often be triggered by other underlying factors. Therefore, the diagnosis of APS relies primarily on the laboratory measurements of aPL. Methods for their determination differ and have not yet been standardized. The common weaknesses of aPL determination are high inter-assay and inter-laboratory variations, problems in interpretation and clinical evaluation of test results, and their low diagnostic specificity. Elevated aPL levels can be associated with many other conditions such as infections, malignancies, and also the use of certain medications. The lack of reliable, robust diagnostic markers for APS thus limits patient identification and treatment and challenges researchers to find better diagnostic markers. A systematic review of observational studies that excluded patients with autoimmune diseases found a pooled prevalence rate of aPL in up to 23.3% of patients with stroke, 23% with myocardial infarction, 15.8% with deep vein thrombosis, and 13% of women with pregnancy adverse events [18].

Many investigators are exploring the usefulness of testing for non-criteria aPL specificities to identify APS in patients with thrombosis and/or pregnancy morbidity, particularly in those who are repeatedly negative on currently used tests. Among them, IgA aPL and antiprothrombin antibodies are most commonly proposed to assess the risk of thrombosis and pregnancy morbidity in patients with suspected APS [19]. A number of studies have shown that antiprothrombin antibodies represent distinct antibody subsets with overall diagnostic relevance for APS [7, 20]. Similar to anti-β2GPI, antiprothrombin antibodies, particularly aPS/PT, have a considerable value as a biomarker for both diagnostic evaluation and prediction of the clinical manifestations of APS. In 2017, a large international multicenter study found that IgG aPS/PT to be more prevalent in patients with APS than in patients without the syndrome. A positive test for these antibodies conferred a 10-fold higher risk of APS [21]. There is debate about the feasibility of including aPS/PT in risk assessment for APS to increase the accuracy of diagnosis in seronegative APS patients [7, 20, 22]. Our research group has extensively studied the clinical significance of antiprothrombin antibodies, showing that aPS/PT have the highest percentage of LA activity compared with aCL or anti-β2GPI [8, 9, 23–26] and that they are strongly associated with thrombosis and adverse pregnancy outcomes independently of other aPL [8, 26]. In fact, aPS/PT were the only antibodies associated with pregnancy complications (recurrent pregnancy loss) occurring before 10-week gestation and with some late complications (preeclampsia and eclampsia), indicating their important role in the pathogenesis of obstetric APS.

Recently, two research groups proposed a quantitative index to quantify the likelihood of thrombosis in APS. One included the aPL profile, the aPL score (aPL-S) [27], whereas the other included both aPL and conventional prothrombotic risk factors, the global APS score (GAPSS) [28]. Both groups included LA and IgG and IgM isotypes of aCL, anti-β2GPI, and aPS/PT. In contrast to risk stratification for thrombotic events, which has been well studied in aPL-positive patients, studies assessing the risk for obstetric complications are scarce. Our recent study investigated different scoring systems after 2 years of routine and systematic measurement of criteria and non-criteria aPL [9]. We showed that all non-criteria aPL, including IgA aCL, IgA anti-β2GPI, and IgA/IgG aPS/PT were as well significantly associated with thrombosis and obstetric complications. We proposed a new quantitative scoring [9] to evaluate the risk of adverse pregnancy events in aPL-positive patients, namely the obstetric risk score—ORS. The ORS showed much higher diagnostic accuracy for obstetric complications compared with any single aPL measure.

4. Thrombotic and obstetric risk assessment

Risk stratification is a major challenge in the management of patients with APS, and a possible role of aPL as a risk or even prognostic factor for arterial/venous thrombosis and miscarriages has been intensively discussed [27, 29]. Single, double, and triple aPL positivity is not uncommon in patients with APS, and such multiple positivity is usually associated with a higher risk for the occurrence or recurrence of thrombotic or obstetric adverse event [30, 31]. Recently, two research groups proposed a quantitative index to quantify the likelihood of thrombosis in APS. One included the aPL profile, and the aPL score (aPL-S) [27], whereas the other included both aPL and conventional prothrombotic risk factors, the global APS score (GAPSS) [28]. Both groups included LA and IgG and IgM isotypes of aCL, anti-β2GPI and aPS/PT.

In contrast to risk stratification for thrombotic events, which has been well studied in aPL-positive patients, studies assessing the risk for obstetric complications are scarce. A recent study examined different scoring systems after 2 years of systematic review [9]. They showed that all non-criteria aPL, including IgA aCL, IgA anti-β2GPI, and IgA/IgG aPS/PT were significantly associated with both thrombosis and obstetric complications. They proposed a novel quantitative scoring to evaluate the risk of adverse pregnancy events in aPL-positive patients, namely the obstetric risk score—ORS. The ORS showed much higher diagnostic accuracy for obstetric complications compared with any single aPL measure.

5. Antiphospholipid antibodies in infections

It is known that aPL may be transiently elevated in sera during various infections, including skin infections (18%), human immunodeficiency virus infections (17%), pneumonia (14%), hepatitis C virus (13%), and urinary tract infections (10%) [32]. The presence of aPL in sera and also its clinical significance was first noted in patients with Treponema pallidum infection [33]. With the continued use of cardiolipin-based serologic tests for syphilis diagnosis, it became apparent that a small group of patients with autoimmune diseases, especially SLE, had "false-positive" tests. In 1983, researchers recognized that the presence of aPL in SLE patients was associated with thromboembolic events and recurrent miscarriage, and the term anticardiolipin syndrome and later antiphospholipid syndrome (APS) were coined [34, 35].

Since the global COVID-19 pandemic, a possible link between the presence of aPL and infection with the SARS-CoV-2 virus has been investigated. Several groups have reported the presence of aPL in patients with COVID-19 and have suggested the possibility of SARS-CoV-2 virus-induced APS [36–38]. Coagulopathy and thrombotic events, including deep vein thrombosis, pulmonary embolism, and stroke, are serious manifestations in critically ill patients with COVID-19.

Currently, the role of aPL in thrombotic complications in COVID-19 is still unclear. Similar to the severe coagulopathies associated with COVID-19, patients with CAPS may develop thrombosis in multiple organs within a very short period of time [39]. Because of the similarity between the course of COVID-19 and CAPS, it was hypothesized that SARS-CoV-2 infection could be a possible trigger for APS. Detailed analysis of 23 studies (with a total of 250 patients) of aPL at COVID-19 showed that the presence of LA, aCL, and anti-β2GPI was 64%, 9%, and 13%, respectively [40]. However, none of the included studies reported re-examination of aPL after 12 weeks, so it is not clear whether the aPL presence in COVID-19 patients was transient or persistent. The only study in which aPL testing was repeated after 1 month and in which aPS/PT was also measured included 31 patients with COVID-19 [41]. In this study, elevated aPL levels were confirmed in 74% of patients, but 9/10 of the LA-positive patients retested were negative the second time. This observation supports the frequent single LA positivity during the acute phase of COVID-19 infection.

Later in the pandemic, two independent reviews were published that examined the prevalence of aPL in COVID-19 patients and its clinical significance [42, 43]. The prevalence of LA ranged from 35 to 92% in ICU patients, aCL IgG in 52%, and IgM in 40% of patients, and anti-β2GPI IgG and IgM were found in up to 39% and up to 34% of patients, respectively. Between 1 and 12% of patients had a triple-positive aPL profile [43]. In the second review, the authors primarily examined studies of aCL and anti-β2GPI but also addressed non-criteria aPL [42]. They concluded that aPL positivity may be a feature of COVID-19, at least in some patients, but in general the identified "solid-phase" aPL are of low titer and cannot be well associated with the thrombotic aspects of COVID-19. Also, in the few studies in which persistence was examined, the results seemed to indicate transient positivity of aPL that occurred only during infection. Importantly, high-titer aPL or multiple APL positivity (including double and triple positivity) was in the minority for COVID-19. There is also one important study where antigen specificity of aPL in COVID-19 has been investigated. These researchers have found that, contrary to APS, which is characterized by high aPL titers with specificity against domain 1 on β2GPI, patients with COVID-19 exhibit low titers of anti-β2GPI, with specificity against domains 4 and 5 [44].

The risk of a recurrent thrombotic event in patients with APS is greatly increased in those who have multiple subtypes of aPL (LA, aCL, anti-β2-GPI, aPS/PT), that is, double-, triple-positive patients. In patients with COVID-19, double or triple aPL positivity appears to be rare and aPL positivity appears to be transient. A well-designed, age- and sex-controlled observational study compared the aPL profile of hospitalized COVID patients with that of a) patients with thrombotic APS and b) patients with culturally/serologically proven infections [45]. Their data showed that positive aPL values can be found in half of the patients with infections, as 53% of patients with COVID-19 and 49% of patients with other viral/bacterial infections had positive aPL values. Importantly, however, the aPL profile was different when comparing patients with overt APS and patients with aPL detected in the setting of infections. Therefore, author conclude, caution is required in interpreting and generalizing the role of aPLs in the management of patients with COVID-19.

6. Antiphospholipid antibodies in cancer

The relationship between thrombosis and cancer was first established by Trousseau in 1865. Since then, numerous studies have shown that thromboembolism is a common complication of cancer, occurring in 15% of all cancer patients [46, 47]. Despite extensive research and modern interventions, thromboembolic disorders are still a major cause of morbidity and mortality in these patients. The risk of thromboembolic events is four times higher in cancer patients than in the general population and this risk is further increased in patients undergoing chemotherapy [47, 48]. Much of this high risk is attributed to the cancer itself. However, patient-related factors such as age, performance status, body mass index, underlying comorbidities, and therapy are also the important factors. The biological origin of thromboembolic events is related to the pro-coagulant, hypoxic, and inflammatory state associated with tumors, especially in advanced stages [49]. Several mechanisms contribute to the hypercoagulable state observed in cancer, resulting in a complex interplay of various factors, including tissue factors, platelet and endothelial activation, coagulation abnormalities, procoagulants secreted by tumor cells, abnormal blood flow, and abnormal tumor angiogenesis [46, 50]. The question arises whether the presence of aPL further increases the thromboembolic risk in patients with malignancies.

A high prevalence of aCL, anti-β2GPI, LA, anti-phosphatidylcholine, anti-phosphatidylserine, anti-phosphatidylinositol, anti-phosphatidylethanolamine, and anti-prothrombin antibodies has been observed in patients with various types of hematologic malignancies and solid tumors [47]. Therefore, the already increased risk of thrombosis in cancer patients is even higher for carriers of aPL. The reported prevalence of elevated aPL levels in cancer patients varies from less than 5%, which is similar to the prevalence observed in healthy individuals, to as high as 70% [47]. This dramatic range is due in part to different methods being performed, differences in study design, and inconsistent definitions of aPL positivity in the medical literature. In general, aPL tests are highly heterogeneous and poorly standardized. In addition, most studies examined the prevalence of aPL only once and did not repeat the test after 3 months, so the frequency may be overestimated. A recent systematic review of observational studies found an increased risk of developing aPL in patients with gastrointestinal, genitourinary, and lung cancer, leading to thromboembolic events and death [51]. In addition, a 17-year observational study of 1592 non-thrombotic women with three consecutive spontaneous abortions before the 10-week gestation or fetal death at or after 10-week gestation showed that the risk of cancer was significantly higher in women with a history of obstetric APS than in the general population [52]. Recently, one research group investigated the presence of criteria and non-criteria aPL in patients with uterine malignancies [53]. The authors found that non-criteria aPL (against phosphatidic acid, phosphatidylserine, annexin V, and prothrombin) are more common in patients with uterine malignancies (UM) than in patients with non-cancerous gynecological diseases (NCGD). In contrast, the criteria aPL did not differ significantly between UM and the NCGD group. It is interesting to note that several studies associate non-criteria aPL, especially antiprothrombin antibodies, with obstetric complications, while they could not confirm the association with either anti-β2GPI or LA [11, 12].

In conclusion, aPL levels appear to be elevated in patients with various malignancies, increasing their risk for thromboembolic events. In the future, it would be important to conduct well-designed large-scale population studies as well as longitudinal studies on patients with various cancers to determine the true risk and confirm whether the increased prevalence of aPL positivity is transient. Although aPL positivity may help assess the risk of blood clots, there are currently no strong data to recommend aPL screening in cancer patients.

7. Antiphospholipid antibodies in healthy individuals

Low aCL levels are found in up to 10% of healthy individuals, and the prevalence of a positive aPL test increases with age [10]. High aPL levels and persistent positivity are rare in healthy individuals (less than 1%). There are no recent studies investigating the level of criterion-related or non-criterion-related aPL in the general population. The clinical significance of aPL in healthy individuals remains unclear. It is important to emphasize that not every positive test for aPL is of clinical significance, and patients with aPL are at different risk for adverse events related to aPL. A rare prospective study in which healthy blood donors were tested for aPL twice 1 year apart showed 10% positivity for aCL and 1% positivity LA at the first measurement. Of note, less than 1% of subjects were still positive after 1 year [54]. Therefore, in parallel with other cardiovascular risk factors such as hypertension, elevated cholesterol, diabetes, smoking or obesity, patients with aPL have a higher risk of adverse events. It is known that aPL can occur transiently during infections or other occasions. This is an important reason why aPL should be tested twice within 12 weeks, which is also embodied in the international classification criteria for APS.

Recently, an administrative database study of aPL in the general population was published that characterized patterns of aPL testing in a sample from the United States using laboratory data from 2010 to 2015. They identified 33,456 individuals with at least one aPL test. Of these, only 6391 (19%) had all three tests (LA, aCL, aGP1) performed. Confirmatory aPL tests were performed at least 12 weeks later in 77, 45, and 41% of initially positive LA, aCL, and aGP1, respectively. Of those retested, only 255 (10.6%) had a confirmatory positive aPL test. The most important finding is the low rate of a confirmatory positive aPL test ≥12 weeks after the first test, indicating that aPL testing is often be incomplete. Further investigation in the form of large-scale population studies as well as longitudinal studies is needed to better understand the clinical relevance of aPL in healthy individuals from different backgrounds.

8. Conclusion

The heterogeneity and non-specificity of the possible clinical symptoms highlight that APS is a true systemic autoimmune disease and emphasizes the need for a better understanding of the disease mechanisms that will allow a personalized treatment approach. In the general population, the incidence of clinical manifestations in APS is high and could often be triggered by other underlying factors. Therefore, the diagnosis of APS relies predominantly on laboratory measurements. Despite the many attempts to increase the specificity of laboratory criteria and to establish consensus criteria for serology, a high number of patients are still misdiagnosed. Treatment of APS requires an interprofessional team approach involving multiple specialties. Family physicians play an important role in identifying patients with APLS. Hematologists and rheumatologists play a critical role in diagnosis, treatment, and follow-up. Involvement of other specialties such as neurology, nephrology, cardiology, and dermatology may also be necessary if a particular organ system is affected. In addition, anticoagulation clinics can play an important role in monitoring therapeutic warfarin levels and INR levels with close follow-up. Last but not the least, pharmacists can help in the management of these patients, especially in identifying drug–drug interactions. Close communication between the interprofessional team and close monitoring of the patient is essential in the management of APS.

Author details

Polona Žigon[1,2]

1 Department of Rheumatology, University Medical Centre Ljubljana, Ljubljana, Slovenia

2 Faculty of Mathematics, Natural Sciences and Information Technologies, University of Primorska, Koper, Slovenia

*Address all correspondence to: polona.zigon@guest.arnes.si

IntechOpen

© 2022 The Author(s). Licensee IntechOpen. This chapter is distributed under the terms of the Creative Commons Attribution License (http://creativecommons.org/licenses/by/3.0), which permits unrestricted use, distribution, and reproduction in any medium, provided the original work is properly cited. CC BY

References

[1] Miyakis S, Lockshin MD, Atsumi T, Branch DW, Brey RL, Cervera R, et al. International consensus statement on an update of the classification criteria for definite antiphospholipid syndrome (APS). Journal of Thrombosis and Haemostasis. 2006;**4**(2):295-306

[2] Schreiber K, Sciascia S, de Groot PG, Devreese K, Jacobsen S, Ruiz-Irastorza G, et al. Antiphospholipid syndrome. Nature Reviews Disease Primers. 2018;**4**:18005

[3] Thomas RH. Hypercoagulability syndromes. Archives of Internal Medicine. 2001;**161**(20):2433-2439

[4] Marchetti T, Cohen M, de Moerloose P. Obstetrical antiphospholipid syndrome: From the pathogenesis to the clinical and therapeutic implications. Clinical & Developmental Immunology. 2013;**2013**:159124

[5] Abreu MM, Danowski A, Wahl DG, Amigo M-C, Tektonidou M, Pacheco MS, et al. The relevance of "non-criteria" clinical manifestations of antiphospholipid syndrome: 14th International Congress on Antiphospholipid Antibodies Technical Task Force Report on Antiphospholipid Syndrome Clinical Features. Autoimmunity Reviews. 2015;**14**(5): 401-414

[6] Soybilgic A, Avcin T. Pediatric APS: State of the art. Current Rheumatology Reports. 2020;**22**(3):9

[7] Sciascia S, Sanna G, Murru V, Roccatello D, Khamashta MA, Bertolaccini ML. Anti-prothrombin (aPT) and anti-phosphatidylserine/prothrombin (aPS/PT) antibodies and the risk of thrombosis in the antiphospholipid syndrome. A systematic review. Thrombosis and Haemostasis. 2013;**111**(2):354-364

[8] Zigon P, Cucnik S, Ambrozic A, Kveder T, Semrl SS, Rozman B, et al. Detection of antiphosphatidylserine/prothrombin antibodies and their potential diagnostic value. Clinical and Developmental Immunology. 2013;**2013**:8

[9] Zigon P, Podovsovnik A, Ambrozic A, Tomsic M, Hocevar A, Gaspersic N, et al. Added value of non-criteria antiphospholipid antibodies for antiphospholipid syndrome: Lessons learned from year-long routine measurements. Clinical Rheumatology. 2018;**38**:371-378

[10] Levine JS, Branch DW, Rauch J. The antiphospholipid syndrome. The New England Journal of Medicine. 2002;**346**(10):752-763

[11] Tektonidou MG, Andreoli L, Limper M, Amoura Z, Cervera R, Costedoat-Chalumeau N, et al. EULAR recommendations for the management of antiphospholipid syndrome in adults. Annals of the Rheumatic Diseases. 2019;**78**(10):1296-1304

[12] Garcia D, Erkan D. Diagnosis and Management of the Antiphospholipid Syndrome. The New England Journal of Medicine. 2018;**378**(21):2010-2021

[13] Asherson RA, Cervera R, Piette JC, Font J, Lie JT, Burcoglu A, et al. Catastrophic antiphospholipid syndrome. Clinical and laboratory features of 50 patients. Medicine. 1998;**77**(3):195-207

[14] Asherson RA, Espinosa G, Cervera R, Font J, Reverter JC. Catastrophic antiphospholipid syndrome: Proposed guidelines for diagnosis and treatment. Journal of Clinical Rheumatology : Practical Reports on Rheumatic & Musculoskeletal Diseases. 2002;**8**(3): 157-165

[15] Kitchens CS. Thrombotic Storm: When Thrombosis Begets Thrombosis. The American Journal of Medicine. 1998;**104**(4):381-385

[16] Ortel TL, Erkan D, Kitchens CS. How I treat catastrophic thrombotic syndromes. Blood. 2015;**126**(11): 1285-1293

[17] Rodríguez-Pintó I, Moitinho M, Santacreu I, Shoenfeld Y, Erkan D, Espinosa G, et al. Catastrophic antiphospholipid syndrome (CAPS): Descriptive analysis of 500 patients from the International CAPS Registry. Autoimmunity Reviews. 2016;**15**(12): 1120-1124

[18] Andreoli L, Chighizola CB, Banzato A, Pons-Estel GJ, Ramire de Jesus G, Erkan D. Estimated frequency of antiphospholipid antibodies in patients with pregnancy morbidity, stroke, myocardial infarction, and deep vein thrombosis: A critical review of the literature. Arthritis Care & Research. 2013;**65**(11):1869-1873

[19] Bertolaccini ML, Amengual O, Atsumi T, Binder WL, de Laat B, Forastiero R, et al. 'Non-criteria' aPL tests: report of a task force and preconference workshop at the 13th International Congress on Antiphospholipid Antibodies, Galveston, TX, USA, April 2010. Lupus. 2011;**20**(2):191-205

[20] Peterson LK, Willis R, Harris EN, Branch WD, Tebo AE. Antibodies to phosphatidylserine/prothrombin complex in antiphospholipid syndrome: analytical and clinical perspectives. In: Makowski GS, editor. Advances in Clinical Chemistry. Vol. 73. Elsevier; 2016. pp. 1-28. ISSN 0065-2423, ISBN 9780128046906 DOI: 10.1016/bs.acc. 2015.10.003. Available from: https://www.sciencedirect.com/science/article/pii/S0065242315000955

[21] Amengual O, Forastiero R, Sugiura-Ogasawara M, Otomo K, Oku K, Favas C, et al. Evaluation of phosphatidylserine-dependent antiprothrombin antibody testing for the diagnosis of antiphospholipid syndrome: Results of an international multicentre study. Lupus. 2017;**26**(3):266-276

[22] Shi H, Zheng H, Yin YF, Hu QY, Teng JL, Sun Y, et al. Antiphosphatidylserine/prothrombin antibodies (aPS/PT) as potential diagnostic markers and risk predictors of venous thrombosis and obstetric complications in antiphospholipid syndrome. Clinical Chemistry and Laboratory Medicine. 2018;**56**(4): 614-624

[23] Zigon P, Ambrozic A, Cucnik S, Kveder T, Rozman B, Bozic B. Modified phosphatidylserine-dependent antiprothrombin ELISA enables identification of patients negative for other antiphospholipid antibodies and also detects low avidity antibodies. Clinical Chemistry and Laboratory Medicine. 2011;**49**(9):1573

[24] Zigon P, Ambrozic A, Mali P, Tomsic M, Sodin Semrl S, Cucnik S. The prevalence and clinical significance of Iga anti-phosphatidylserine/prothrombin antibodies in systemic autoimmune diseases. Immunome Research. 2017;**13**:130

[25] Žigon P, Božič-Mijovski M, Frank Bertoncelj M, Ambrožič A, Tomšič M, Hočevar A, et al. Laboratory methodology important in the diagnosis and prognosis of antiphospholipid syndrome. In: Bozic-Mijovski M, editor. Thrombosis, Atherosclerosis and Atherothrombosis - New Insights and Experimental Protocols. Rijeka, Croatia: InTech; 2015. Available from: http://www.intechopen.com/books/thrombosis-atherosclerosis-and-atherothrombosis-new-insights-and-experimental-protocols/laboratory-methodology-important-in-the-diagnosis-and-prognosis-of-antiphospholipid-syndrome

[26] Zigon P, Perdan Pirkmajer K, Tomsic M, Kveder T, Bozic B, Sodin Semrl S, et al. Anti-phosphatidylserine/prothrombin antibodies are associated with adverse pregnancy outcomes. Journal of Immunology Research. 2015;**2015**:975704

[27] Otomo K, Atsumi T, Amengual O, Fujieda Y, Kato M, Oku K, et al. Efficacy of the antiphospholipid score for the diagnosis of antiphospholipid syndrome and its predictive value for thrombotic events. Arthritis and Rheumatism. 2012;**64**(2):504-512

[28] Sciascia S, Sanna G, Murru V, Roccatello D, Khamashta MA, Bertolaccini ML. GAPSS: The Global Anti-Phospholipid Syndrome Score. Rheumatology (Oxford, England). 2013;**52**(8):1397-1403

[29] Andreoli L, Tincani A. Beyond the "syndrome": antiphospholipid antibodies as risk factors. Arthritis and Rheumatism. 2012;**64**(2):342-345

[30] Hernandez-Molina G, Espericueta-Arriola G, Cabral AR. The role of lupus anticoagulant and triple marker positivity as risk factors for rethrombosis in patients with primary antiphospholipid syndrome. Clinical and Experimental Rheumatology. 2013;**31**(3):382-388

[31] Pengo V, Ruffatti A, Legnani C, Gresele P, Barcellona D, Erba N, et al. Clinical course of high-risk patients diagnosed with antiphospholipid syndrome. Journal of Thrombosis and Haemostasis. 2010;**8**(2):237-242

[32] Cervera R, Asherson RA, Acevedo ML, Gomez-Puerta JA, Espinosa G, De La Red G, et al. Antiphospholipid syndrome associated with infections: clinical and microbiological characteristics of 100 patients. Annals of the Rheumatic Diseases. 2004;**63**(10):1312-1317

[33] Pangborn MC. A new serologically active phospholipid from beef heart. Proceedings of the Society for Experimental Biology and Medicine. 1941;**48**(2):484-486

[34] Harris EN, Gharavi AE, Boey ML, Patel BM, Mackworth-Young CG, Loizou S, et al. Anticardiolipin antibodies: Detection by radioimmunoassay and association with thrombosis in systemic lupus erythematosus. Lancet. 1983;**2**(8361): 1211-1214

[35] Hughes GR. The anticardiolipin syndrome. Clinical and Experimental Rheumatology. 1985;**3**(4):285-286

[36] Cavalli E, Bramanti A, Ciurleo R, Tchorbanov AI, Giordano A, Fagone P, et al. Entangling COVID-19 associated thrombosis into a secondary antiphospholipid antibody syndrome: Diagnostic and therapeutic perspectives (Review). International Journal of Molecular Medicine. 2020;**46**(3): 903-912

[37] Ehrenfeld M, Tincani A, Andreoli L, Cattalini M, Greenbaum A, Kanduc D, et al. Covid-19 and autoimmunity. Autoimmunity Reviews. 2020;**19**(8):102597

[38] Marietta M, Coluccio V, Luppi M. COVID-19, coagulopathy and venous thromboembolism: more questions than answers. Internal and Emergency Medicine. 2020;**15**(8):1375-1387

[39] Cárdenas Suri H, Jimomila BD. Catastrophic antiphospholipid antibody syndrome and multiple organ dysfunctions in critically ill patients with COVID-19. Expert Review of Respiratory Medicine. 2020;**14**(11):1071-1072

[40] Gkrouzman E, Barbhaiya M, Erkan D, Lockshin MD. A reality check on antiphospholipid antibodies in COVID-19-associated coagulopathy. Arthritis & Rheumatology. 2021;**73**(1):173-174

[41] Devreese KMJ, Linskens EA, Benoit D, Peperstraete H. Antiphospholipid antibodies in patients with COVID-19: A relevant observation? Journal of Thrombosis and Haemostasis. 2020;**18**:2191-2201

[42] Favaloro EJ, Henry BM, Lippi G. COVID-19 and antiphospholipid antibodies: Time for a reality check? Seminars in Thrombosis and Hemostasis. 2022;**48**(1):72-92

[43] Foret T, Dufrost V, Salomon Du Mont L, Costa P, Lefevre B, Lacolley P, et al. Systematic review of antiphospholipid antibodies in COVID-19 patients: Culprits or bystanders? Current Rheumatology Reports. 2021;**23**(8):65

[44] Borghi MO, Beltagy A, Garrafa E, Curreli D, Cecchini G, Bodio C, et al. Anti-phospholipid antibodies in COVID-19 are different from those detectable in the anti-phospholipid syndrome. Frontiers in Immunology. 2020;**11**:584241

[45] Sciascia S, Radin M, Bazzan M, Montaruli B, Cosseddu D, Norbiato C, et al. Antiphospholipid antibodies and infection: Non Nova Sed Nove. Frontiers in Immunology. 2021;**12**: 687534

[46] Reinstein E, Shoenfeld Y. Antiphospholipid syndrome and cancer. Clinical Reviews in Allergy & Immunology. 2007;**32**(2):184-187

[47] Islam MA. Antiphospholipid antibodies and antiphospholipid syndrome in cancer: Uninvited guests in troubled times. Seminars in Cancer Biology. 2020;**64**:108-113

[48] Khorana AA, Francis CW, Culakova E, Kuderer NM, Lyman GH. Frequency, risk factors, and trends for venous thromboembolism among hospitalized cancer patients. Cancer. 2007;**110**(10):2339-2346

[49] Solinas C, Saba L, Sganzerla P, Petrelli F. Venous and arterial thromboembolic events with immune checkpoint inhibitors: A systematic review. Thrombosis Research. 2020;**196**:444-453

[50] Mukai M, Oka T. Mechanism and management of cancer-associated thrombosis. Journal of Cardiology. 2018;**72**(2):89-93

[51] Abdel-Wahab N, Tayar JH, Fa'ak F, Sharma G, Lopez-Olivo MA, Yousif A, et al. Systematic review of observational studies reporting antiphospholipid antibodies in patients with solid tumors. Blood Advances. 2020;**4**(8):1746-1755

[52] Gris J-C, Mousty É, Bouvier S, Ripart S, Cochery-Nouvellon É, Fabbro-Peray P, et al. Increased incidence of cancer in the follow-up of obstetric antiphospholipid syndrome within the NOH-APS cohort. Haematologica. 2020;**105**(2):490-497

[53] Majdan A, Majdan M, Dryglewska M, Ziober-Malinowska P, Kotarski J, Grzybowska-Szatkowska L. The presence of particular criteria and non-criteria antiphospholipid antibodies in patients with uterine malignancies. Polish Archives of Internal Medicine. 2020;**130**:1037-1042

[54] Ünlü O, Zuily S, Erkan D. The clinical significance of antiphospholipid antibodies in systemic lupus erythematosus. European Journal of Rheumatology. 2016;**3**(2):75-84

Section 2

Clinical Aspects of the Antiphospholipid Syndrome

Chapter 2

Antiphospholipid Syndrome and Stroke

Kathryn Grimes, Adam P. Klein, Rakhee Lalla, Adeolu Morawo, Sana Somani, Matthew J. Woodward and John W. Cole

Abstract

Thromboses of the cerebral arterial and venous systems are a common manifestation of antiphospholipid syndrome (APS) often leading to ischemic and hemorrhagic stroke. APS increases stroke risk via many mechanisms, including hypercoagulability and inflammation. These mechanisms, among others, must be considered by physicians when evaluating and treating such patients to achieve optimal short- and long-term outcomes. In this chapter, we will discuss the epidemiology of APS as it relates to neurological disease focusing on stroke, APS stroke mechanisms, suggested clinical evaluations, acute treatment strategies, and long-term secondary stroke prevention strategies. Current consensus statements and the most recent literature will be summarized.

Keywords: antiphospholipid syndrome, stroke, epidemiology, etiology, treatment

1. Introduction

Antiphospholipid syndrome (APS) was first described in 1983 with steadily improving clinical and scientific refinements since that time. It was initially recognized with the discovery of lupus anticoagulant immunoglobulin that binds to phospholipids and proteins associated with the cell membrane and its association with other autoimmune conditions. Over the years, the clinical manifestations of APS were further delineated, followed by the discovery of other antiphospholipid antibodies. Currently, APS is defined as an autoimmune condition characterized by the presence of venous or arterial thrombosis and/or pregnancy-related complications in patients with antiphospholipid antibodies [1]. Notably, APS can occur as a *primary* disease process or *secondary* to another condition, primarily autoimmune conditions, including systemic lupus erythematous (SLE), rheumatoid arthritis, sjogren's disease, or systemic sclerosis. It can more rarely be secondary to malignancy [2] and infections, including syphilis and HIV [3].

Clinically, APS can manifest in a variety of ways and affect multiple organ systems. Presenting symptoms can range from relatively benign to severe. One subtype (to be discussed in Section 2) termed catastrophic APS (CAPS) is defined as APS that affects >3 organs in a short period of time (<7 days) with pathologic evidence of small-vessel occlusion. The most common venous manifestation of APS is deep vein thrombosis, while stroke is the most common arterial manifestation of this disease [4]. Obstetric complications include placental insufficiency and recurrent pregnancy loss, typically after 10 weeks of gestation. There are, however, a multitude of other manifestations including cardiac valvular disease, coronary artery disease, livedo reticularis, renal small artery vasculopathy, and thrombocytopenia, which are *not* included in the formal classification criteria [1]. Neurologically, antiphospholipid antibodies have also been found to be more rarely related to migraine, seizures, movement disorders, and cognitive impairment [5]. Given this broad range of clinical manifestations, it is important that clinicians have a clear understanding of when to suspect this condition and its appropriate management.

Antiphospholipid antibodies (aPL) are a serological marker for APS and their presence is key to the definition and classification for APS. Phospholipids are molecules found in the blood that aid in clot formation. They form complexes with other plasma proteins and are the target of aPL antibodies; thus, one may expect to clinically see a bleeding disorder when phospholipids are disrupted. However, these autoantibodies primarily cause endothelial dysfunction and disruption of coagulation factors as they compete with coagulation factors for available phospholipids, thereby leading to a procoagulant state and clot formation [6]. The pathophysiology of aPL antibodies is not fully elucidated, but the current thought is that of a "two-hit" hypothesis. The first hit being a patient-specific susceptibility, and the second hit being a trigger or inciting event. This theory is based on the idea that about 1–5% of the population may have positive aPL antibodies without any clinical manifestations, indicating the need for a trigger that leads to the pathologic state [2, 4]. In a patient carrying aPL antibodies, endothelial cell activation occurs in the setting of oxidative stress in conditions such as infection, surgery, and pregnancy. This is thought to subsequently lead to a series of events including complement activation, cytokine release, increased expression of tissue factor on endothelial cells, increased platelet adhesiveness, and impairment of thrombolysis [2, 4]. Overall, this creates a procoagulant state leading to the range of clinical manifestations as described.

aPL antibodies are a heterogeneous group of autoantibodies that primarily include *lupus anticoagulant (LA), anti-cardiolipin IgG/IgM (aCL), and anti-beta-2 glycoprotein-I (aB2GPI) IgG/IgM,* with these three specific antibodies included in the formal classification criteria for APS [1]. As shown in **Figure 1** there is some overlap between these antibodies, but overall, they are distinct leading to a variety of clinical manifestations [5]. In addition to the three antibodies in the classification criteria, there are a number of other proposed antibodies of yet unclear clinical significance and diagnostic value. These include anti-prothrombin and anti-phosphatidylserine-prothrombin complex, aCL IgA and anti-B2GPI IgA. These antibodies are sometimes used to aid in diagnosis if there is a very high clinical suspicion for APS without the presence of the typical autoantibodies in the classification criteria [7]. It is important to note that while B2GPI is considered a primary APS antigen, subgroups of protein domains can be targeted by specific antibodies. For example, antibodies targeting B2GPI Domain I, in particular, have been correlated with a high risk of thrombosis [8].

The presence of LA alone is thought to hold the highest risk for thrombosis among all antiphospholipid antibodies. Thrombotic risk is much lower in patients who have only a positive aCL or anti-B2GPI antibody [1, 3]. The risk is thought to be much higher however in patients with multiple positive antibodies, especially those found to be "triple positive" [3]. Thrombotic risk is also much higher in patients who have secondary APS is associated with SLE and in patients with primary APS with concurrent vascular comorbidities including hypertension, hypercholesterolemia, tobacco, and oral contraceptive use [7].

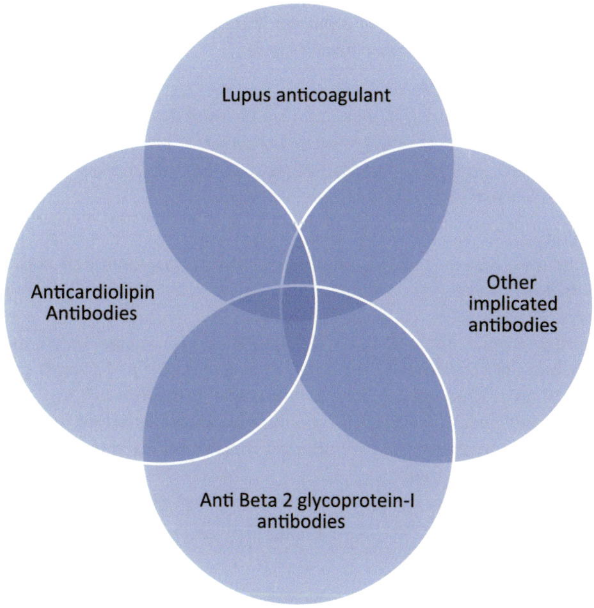

Figure 1.
There are a variety of antiphospholipid antibodies associated with APS, as detected with different methods, some are overlapping, but each has distinct properties. Image adapted from Misita et al. [6].

The initial classification criteria for APS, called the Sapporo criteria, was first developed in 1999 and most recently updated in 2006 [1]. As shown in **Table 1**, the criteria currently require one clinical manifestation of thrombosis or pregnancy complication, and one laboratory criteria present on two occasions at least 12 weeks apart.

As mentioned, there are other autoantibodies implicated in APS that are not yet included in the classification criteria. The remainder of this chapter will discuss the clinical manifestations, epidemiology, pathophysiology, diagnosis, and treatment in more detail.

2. Clinical presentation

APS can present as a wide range of clinical manifestations with the major clinical features consisting of arterial and venous thromboses, and obstetrical complications. The most common obstetrical manifestations of APS are recurrent early miscarriage, placental insufficiency, early pre-eclampsia, and fetal death, all of which should prompt evaluation for the presence of aPL [12].

Thrombotic events in APS may occur in virtually any vascular bed, with the cerebral circulation being the arterial territory most commonly affected, usually in the form of stroke or transient ischemic attack [13]. APS has also been associated with many other clinical features including livedo reticularis, epilepsy, thrombocytopenia, and cognitive dysfunction, however, the strength of association is not sufficiently high to include them in the syndrome definition. The clinical characteristics of a cohort of 1000 patients with APS (Euro-Phospholipid Project) are displayed in **Table 2** [14].

2.1 Classification criteria: additional considerations

As described in Section 1, the first set of criteria for APS was established in Sapporo, Japan in 1999 after an expert workshop [9]. This was modified, including

Antiphospholipid antibody syndrome (APS) is present if at least one of the clinical criteria and one of the laboratory criteria that follow arc met[*] clinical criteria

1. Vascular thrombosis[†]
One or more clinical episodes[‡] of arterial, venous, or small vessel thrombosis[§], in any tissue or organ. Thrombosis must be confirmed by objective validated criteria (i.e. unequivocal findings of appropriate imaging studies or histopathology). For histopathologic confirmation, thrombosis should be present without significant evidence of inflammation in the vessel wall.

2. Pregnancy morbidity
   a. One or more unexplained deaths of a morphologically normal fetus at or beyond the 10th week of gestation, with normal fetal morphology documented by ultrasound or by direct examination of the fetus, or
   b. One or more premature births of a morphologically normal neonate before the 34th week of gestation because of (i) eclampsia or severe pre-eclampsia defined according to standard definitions [9], or (ii) recognized features of placental insufficiency[¶], or
   c. Three or more unexplained consecutive spontaneous abortions before the 10th week of gestation, with maternal anatomic or hormonal abnormalities and paternal and maternal chromosomal causes excluded.

In studies of populations of patients who have more than one type of pregnancy morbidity, investigators arc strongly encouraged to stratify groups of subjects according to a, b, or c above.

Laboratory criteria[**]
1. Lupus anticoagulant (LA) present in plasma, on two or more occasions at least 12 weeks apart, detected according to the guidelines of the International Society on Thrombosis and Hemostasis (Scientific Subcommittee on LAs/phospholipid-dependent antibodies) [10, 11].

2. Anticardiolipin (aCL) antibody of IgG and/or IgM isotype in scrum or plasma, present in medium or high titer (i.e. >40 GPL or MPL, or >the 99th percentile), on two or more occasions, at least 12 weeks apart, measured by a standardized ELISA.

3. Anti-β_2 glycoprotein-I antibody of IgG and/or IgM isotype in scrum or plasma (in titer > the 99th percentile), present on two or more occasions, at least 12 weeks apart, measured by a standardized ELISA, according to recommended procedures.

[*]Classification of APS should be avoided if less than 12 weeks or more than 5 years separate the positive aPL test and the clinical manifestation.

[†]Coexisting inherited or acquired factors for thrombosis arc, not reasons for excluding patients from APS trials. However, two subgroups of APS patients should be recognized, according to (a) the presence, and (b) the absence of additional risk factors for thrombosis. Indicative (but not exhaustive) such eases include: age (>55 in men, and >65 in women), and the presence of any of the established risk factors for cardiovascular disease (hypertension, diabetes mellitus, elevated LDL or low HDL cholesterol, cigarette smoking, family history of premature cardiovascular disease, body mass index ≥30 kg m^{-2}, microalbuminuria, estimated GFR < 60 ml min^{-1}), inherited thrombophilias, oral contraceptives, nephrotic syndrome, malignancy, immobilization, and surgery. Thus, patients who fulfill criteria should be stratified according to contributing causes of thrombosis.

[‡]A thrombotic episode in the past could be considered as a clinical criterion, provided that thrombosis is proved by appropriate diagnostic means and that no alternative diagnosis or cause of thrombosis is found.

[§]Superficial venous thrombosis is not included in the clinical criteria.

[¶]Generally accepted features of placental insufficiency include: (i) abnormal or non-reassuring fetal surveillance test (s), e.g. a non-reactive non-stress test, suggestive of fetal hypoxemia, (ii) abnormal Doppler flow velocimetry waveform analysis suggestive of fetal hypoxemia, e.g. absent end-diastolic flow in the umbilical artery, (iii) oligohydramnios, e.g. an amniotic fluid index of 5 cm or less, or (iv) a postnatal birth weight less than the 10th percentile for the gestational age.

[**]Investigators arc strongly advised classifying APS patients in studies into one of the following categories: I, more than one laboratory criteria present (any combination): IIa, LA present alone; IIb, aCL antibody present alone; IIc, anti-β_2 glycoprotein-I antibody present alone.

Table 1.
The classification criteria for APS [1].

the addition of anti-β2GPI antibodies in Sydney, Australia in 2006. The revised APS classification criteria strongly recommend investigating coexisting inherited and acquired thrombosis risk factors in patients with APS [1]. A recent assessment of the 2006 revised APS classification criteria has shown that only 59% of the patients meeting the 1999 APS Sapporo classification criteria met the revised criteria [15]. In addition, many of the older studies evaluated for only a few of the specific aPL

Manifestation	No. (%) of patients
Peripheral thrombosis	
Deep vein thrombosis	389 (38.9%)
Other peripheral thrombi	248 (24.8%)
Neurologic manifestations	
Migraine	202 (20.2%)
Stroke	198 (19.8%)
Transient ischemic attack	111 (11.1%)
Epilepsy	70 (7.0%)
Multi-infarct dementia	25 (2.5%)
Chorea	13 (1.3%)
Acute encephalopathy	11 (1.1%)
Transient amnesia	7 (0.7%)
Cerebral venous thrombosis	7 (0.7%)
Cerebellar ataxia	7 (0.7%)
Transverse myelopathy	4 (0.4%)
Hemiballismus	3 (0.3%)
Pulmonary manifestation	
Pulmonary embolism	141 (14.1%)
Other pulmonary manifestations	56 (5.6%)
Cardiac manifestations	
Valve thickening/dysfunction	116 (11.6%)
Other cardiac manifestations	153 (15.3%)
Intraabdominal manifestations	
Renal manifestations	27 (2.7%)
Gastrointestinal manifestations	42 (4.2%)
Cutaneous manifestations	
Livedo reticularis	241 (24.1%)
Other cutaneous manifestations	155 (15.5%)
Osteoarticular manifestations	
Arthralgia	387 (38.7%)
Other osteoarticular manifestations	295 (29.5%)
Ophthalmological manifestations	
Amaurosis fugax	54 (5.4%)
Other ophthalmological manifestations	34 (3.4%)
Ear, nose, throat manifestations	8 (0.8%)
Hematologic manifestations	
Thrombocytopenia	296 (29.6%)
Hemolytic anemia	97 (9.7%)
Obstetric manifestations (n = 590 pregnant women)	
Preeclampsia	56 (9.5%)

Manifestation	No. (%) of patients
Other obstetric manifestations	41 (7.1%)
Fetal manifestations (n = 1580 pregnancies)	
Live birth	753 (47.7%)
Other fetal manifestations (fetal loss, premature births)	827 (52.3%)

Table 2.
Cumulative clinical features during the evolution of the disease in 1000 patients with APS (adapted [14]).

Criteria
1. Evidence of involvement of three or more organs, systems, and/or tissues.
2. Development of manifestations simultaneously or in less than a week.
3. Confirmation by histopathology of small vessel occlusion in at least one organ or tissue.
4. Laboratory confirmation of the presence of antiphospholipid antibodies (lupus anticoagulant, anticardiolipin antibodies, and/or anti-beta2-glycoprotein I antibodies).
Classification
Definite catastrophic APS
Requires all four criteria
Probable catastrophic APS
All four criteria, except for only two organs, systems, and/or sites of tissue involvement **or**
All four criteria, except for the laboratory confirmation at least six weeks apart due to the early death of a patient never tested for aPL before the catastrophic APS **or**
Criteria 1, 2, and 4 above **or**
1, 3, and 4 and the development of the third event in more than a week but less than a month, despite anticoagulation.

Table 3.
Preliminary criteria for the classification of catastrophic antiphospholipid syndrome (CAPS) [18, 19].

antibodies now thought to be important in stroke risk, accepted low positive titers and many looked at only one-time point, hence it is difficult to apply the results of those studies [16]. While the purpose of the criteria was to help choose patients for clinical trials, it is the best available tool to avoid over-diagnosis of APS in clinical practice [17].

CAPS is a rare and potentially fatal complication of APS. As described in **Table 3**, the clinical presentation is characterized by acute multi-organ failure due to thromboses of three or more organs within 1 week, associated with the presence of aPL and thrombocytopenia [16]. CAPS can be seen as the first presentation of APS or can be triggered by infection, surgery, or trauma in patients with known APS [19].

In the setting of pregnancy, Obstetric APS (OAPS) is diagnosed if at least one of the clinical criteria and one of the laboratory criteria are met as outlined in **Table 4** [1, 20].

2.2 Ischemic stroke

Although up to 5% of the population might be positive for aPL antibodies, only a small fraction is diagnosed with APS as per the mentioned criteria [21]. Based on the analysis of 120 full-text papers, the overall estimated aPL frequency in stroke

Clinical criteria	Laboratory criteria
1. One or more unexplained deaths of a morphologically normal fetus at or beyond the 10th week of gestation. 2. One or more preterm births of a morphologically normal neonate before the 34th week of gestation because of: i. eclampsia or severe pre-eclampsia or ii. recognized features of placental insufficiency. 3. Three or more unexplained consecutive spontaneous miscarriages before the 10th week of gestation, with maternal anatomic or hormonal abnormalities and paternal and maternal chromosomal causes excluded.	1. LA present in plasma, on two or more occasions at least 12 weeks apart. 2. aCL of immunoglobulin (Ig)G and/or IgM isotype in serum or plasma, present in medium or high titer (i.e. >40 GPL units or MPL units, or > the 99th percentile), on two or more occasions, at least 12 weeks apart. 3. Anti-β2GPI of IgG and/or IgM isotype in serum or plasma (in titer > the 99th percentile), present on two or more occasions at least 12 weeks apart.

Table 4.
Obstetric APS (OAPS) is diagnosed if at least one of the clinical criteria and one of the laboratory criteria are met [1, 20].

patients of all ages is 13.5% [22]. Sciascia et al. [7], in a systematic review of data from 5217 patients concluded that the overall aPL frequency was estimated as 17.2% for stroke and 11.7% for the transient ischemic attack, and the presence of aPL seems to confer a five-fold higher risk for stroke or TIA when compared with controls. The cumulative prevalence in the Euro-Phospholipid Project Study was 19.8% for stroke and 11.1% for TIA [14], making it the most common and severe arterial complication of APS.

Notably, it has been suggested that more than 20% of strokes in patients younger than 45 years are associated with APS [23], although this estimate may be inflated by referral bias [24]. The presence and magnitude of the ischemic stroke risk associated with aPL in the older population are more evenly split between finding an increased risk and no increased risk. This suggests that aPL may be a more important stroke mechanism in young people whereas, in older populations, other stroke risk factors take on a greater importance.

aPL associated strokes pose a higher risk for women. The Framingham cohort and offspring study found an increased risk of strokes and TIAs for women with high anticardiolipin but not in men [25]. In another study of 34 women under 45 years of age with ischemic strokes and no traditional vascular risk factors, 35% were found to have anticardiolipin antibodies [26].

Another study demonstrated that high serum concentrations of aPL, regardless of other cardiovascular risk factors, were an important predictor of the risk of future stroke and TIA in only females [27]. The presence of anti-β2GP1 antibodies in young women may increase the stroke risk 2.3-fold according to the RATIO study [28].

In terms of traditional vascular risk factors in APS patients, it is debated whether these or the circulating aPL antibodies are responsible for the accelerated atherosclerosis seen in APS. Hypertension is more prevalent in SLE and APS than in the general population. A study showed that hypertension was the only independent risk factor for arterial manifestations, mainly stroke, in APS [29]. The risk of stroke for LA-positive patients was two-fold in smokers and six-fold in smokers receiving oral contraceptives [25]. The Italian Project on Stroke in Young Adults, a prospective study of 1867 patients showed that family history of strokes, migraines with auras, aPL, discontinuation of antiplatelet or antihypertensive medications and increase in at least one traditional vascular risk factor were independent predictors for thromboembolic events [30]. Overall, this emphasizes the importance of aggressively treating all modifiable stroke risk factors like hypertension, diabetes,

Patient age < 50 years of age
Female gender
Lack of traditional vascular risk factors
Positive family history for arterial or venous thromboses
Recurrent strokes
Thrombocytopenia, obstetric complications, venous thromboses, or other arterial thromboses
SLE or presence of other connective tissue diseases

Table 5.
Key factors warranting evaluation of antiphospholipid syndrome.

hypercholesterolemia, obesity, OCP use, and tobacco use to reduce additional thrombotic risks.

A summary of factors that warrant an evaluation of APS in stroke patients is listed in **Table 5**.

Stroke subtypes in APS may be either thrombotic or cardioembolic depending on the location and size of the occluded vessel [31]. Intracranial stem or branch arterial occlusions and stenosis were reported in 50% of APS patients with stroke [32]. Narrowing of multiple intracranial arteries may occur in APS and indicates vasculopathy rather than vasculitis. Occasionally, there is involvement of the extracranial carotid artery. In a small case series of 17 patients, 32% had extracranial arterial abnormalities [33]. Cardioembolic strokes in APS are associated with left cardiac valvular abnormalities, including irregular thickening of leaflets, non-bacterial vegetations, and valve dysfunction [32]. Stroke subtypes in APS can also vary according to the types of antibodies [34]. Saidi et al. [35], in an analysis of 208 patients with their first stroke, reported that antiphosphatidylserine IgG was associated with cardioembolic strokes, lupus anticoagulant with lacunar strokes, and anticardiolipin IgG and IgM with lacunar, atherosclerotic and cardioembolic strokes. The severity of the thromboembolic event does not relate to the aPL antibody titer.

The type of antibodies present also appears to have an association with increased thrombotic risk. The presence of antiphosphatidylserine antibodies had the highest risk for clinical manifestations of APS, and IgG antiphosphatidylserine antibodies correlated strongly with the presence of lupus anticoagulant. The presence of antiphosphatidylserine antibodies (IgG or IgM) or anti-b2GP-1 (IgG, IgM, or IgA) antibodies improved the specificity for APS over anticardiolipin antibodies alone [36]. In another study, the positive predictive value for antiphosphatidylserine and anti-b2GP-1 antibodies was stronger for arterial thromboses than for venous thromboses [37]. Another study of pregnant women with APS reported that patients with triple aPL positivity (LA, aCL, and anti-B2GPI) and/or previous thromboembolism had an increased likelihood of poor neonatal outcomes than patients with double or single aPL positivity and no thrombosis history [38].

The recurrent risk of stroke in APS patients has been less widely studied as compared to other types of thromboses. Pezzini et al. calculated a cumulative risk of 14% for brain ischemia at 10 years [30]. Recurrent strokes and other thromboembolic events in patients with aPL antibodies have been reported both early (within the first year of an index stroke event) and late (5–10 years) [39]. The initial type of thromboembolic event (i.e. arterial, venous, miscarriage) appears to be the most likely type of event to recur in a given patient according to some studies [40]. The Euro-Phospholipid Project Group reported thrombotic events in 16.6% of patients in the first 5 years of follow-up and in 14.4% in the second 5-year follow-up period.

Factor	Point value
Anticardiolipin Antibody IgG/IgM	5
Anti-B2-glycoprotein I IgG/IGM	4
Lupus anticoagulant	4
Hyperlipidemia	3
Arterial hypertension	1

Table 6.
Adjusted global antiphospholipid syndrome score. Adapted [41, 42].

The most common events during follow-up were strokes, TIAs, DVTs, and pulmonary emboli with survival probability at 10 years being 90.7% [14].

The first model to develop a predictive model for aPL associated thrombosis risk in SLE patients was modified in 2013 by Sciascia et al. to include data on clinical manifestations, and risk factors forming a quantitative score called the Global Antiphospholipid Syndrome Score (GAPSS) [41]. This was further modified in 2019 to form the aGAPSS (Adjusted Global Antiphospholipid Syndrome Score) as outlined in **Table 6** [42]. The goal of the aGAPSS is to risk-stratify patients based on the likelihood of developing recurrent thrombosis in the setting of APS.

Taken together, screening for APS is indicated in stroke patients who meet even some of the clinical and laboratory criteria and those with recurrent strokes despite maximal medical management and no clear etiology. The goal of these scoring systems is to further refine the risk of recurrent thromboses associated with APS.

2.3 Venous sinus thrombosis

Cerebral venous sinus thrombosis (CVST) usually presents with headaches, nausea, vomiting, often associated with seizures, and focal neurological deficits. Papilledema, coma, and death also occasionally contribute to the clinical manifestation of CVST. In patients with CVST, reported frequency of aCL positivity ranges from 7 to 22% [43], and predisposes to CVST at a relatively younger age and to a more extensive cerebral venous involvement [44]. In addition, a higher rate of post-cerebral venous sinus thrombosis headache and more infarctions on brain imaging studies are seen in patients with aPL antibodies than in those without them [45].

2.4 Other neurologic manifestations

While intracranial hemorrhage (ICH) is not a common manifestation of APS, there have been reports of reversible vasoconstriction syndrome (RCVS) [46] which is characterized by thunderclap headaches (severe pain peaking in seconds), and focal neurologic deficits.

Moyamoya disease, a progressive narrowing of cerebral vasculature with collateralization, has also been reported to have associations with APS. Of the 16 cases reported in a small series of moyamoya and aPL, 21% fulfilled APS criteria [47].

Sneddon syndrome is a rare entity that may be considered during workup for APS. It is a chronic disorder, usually non-inflammatory, notable for generalized livedo racemosa (which may be confused with livedo reticularis seen in APS), and recurrent strokes [48]. Livedo racemosa is characterized by a violaceous netlike patterning of the skin similar to the familiar livedo reticularis, although it differs by its location (more generalized and widespread, found not only on the limbs but also on the trunk and/or buttocks). Approximately 40–50% of patients with Sneddon's syndrome present aPL antibodies, suggesting that some patients should be classified as APS [49].

Cognitive dysfunction has been reported 19–40% in aPL-positive patients [50]. While many believe that the cognitive decline is due to multiple subcortical infarcts, there have been theories that it is multifactorial, with genetic predisposition, antibody specificity, and direct antibody effects as potential contributors [51].

Migraines are the most prevalent neurologic manifestation in APS, estimated prevalence of around 20% [52].

Other rare clinical manifestations of APS include seizures, acute ischemic encephalopathy, transverse myelitis, amaurosis fugax, optic neuropathy, and other neuropsychiatric disorders.

3. Epidemiology of stroke in the setting of APS

3.1 How many strokes can be attributed to antiphospholipid antibodies?

APS has been a recognized cause of cerebrovascular events (CVE) especially in those without classic cardiovascular risk factors. Traditionally, it has been estimated that one in five strokes in patients younger than 45 could be associated with APS, but there have been concerns that this is an over-estimate due to referral bias [53]. Systematic reviews have provided much of our current knowledge on the prevalence of aPL in patients with vascular events, however broad population studies are lacking. One large study evaluating stroke, pregnancy morbidity, myocardial infarction, and deep vein thrombosis estimated that aPL antibodies were present in ~14% of stroke patients [22].

APS, either primary or secondary, garners consideration especially in young patients with CVE. To address events in the young, the previous study [21] was repeated for those less than 50 years of age and positive aPL was found in 17.4% of cases [54]. Regardless of diagnosis, the presence of any aPL increased the risk of CVE by 5.48-fold for those under the age of 50, and the risk of thrombosis progressively increases with the increasing number of positive antibodies [54]. It has also been reported that patients with stroke and aPL positivity are younger and more likely to be female than patients with strokes who are aPL negative [51]. A similar risk for CVE has been recently reported in another study, where persistently positive aPL increased the risk of CVE by 4.62-fold and where the positive criteria and non-criteria aPL was found in 20/89 (22%) CVE patients [55].

3.2 How common are cerebrovascular events among patients with APS?

The Euro-Phospholipid Project cataloged the largest group of patients with APS. At the initiation of this study, prevalence data were obtained with 13.1% of patients having a stroke as their presenting manifestation [52]. Stroke was the fourth most common presenting symptom behind deep vein thrombosis, thrombocytopenia, and livedo reticularis. Of the 1000 patients, 204 (about 20%) experienced a stroke at some point during their disease course [52]. Cervera et al. [52] made a delineation regarding age-of-onset, defining "older-onset" APS as diagnosis after the age of 50. Comparatively, the over-50 patients were more likely to have strokes (30%) and were more likely to be male (34%), and were more likely to experience angina pectoris (9%) [52]. These patients were followed over a 10-year time period, and over that time period, 5.3% of the patients experienced a stroke. Stroke was the most prevalent thrombotic event. It was also the 4th leading cause of death in these patients following bacterial infection, myocardial infarction, and malignancy [14].

Patients with APS hospitalized with a stroke also have increased mortality compared to patients without APS [55]. APS has also been identified as an independent risk factor for hemorrhagic transformation of ischemic stroke (OR 2.57, 95%CI 1.14–5.81, p = 0.0228) and extended hospital length of stay [56].

3.3 What types of cerebrovascular events occur in patients with APS?

One of the unique aspects of APS is the diversity of types of vasculature involved—arteries and veins, small vessels, and large vessels. Multiple mechanisms of the prothrombotic state have been theorized and will be discussed in Section 4 of this chapter. APS has been implicated in multiple stroke etiologic subtypes including large-artery atherosclerosis, cardio-embolism, and small-vessel occlusion. However, the percentage breakdown between these etiologies has not been consistently reported.

As previously stated, APS is responsible for venous events as well as arterial events. In the cerebrovascular system, these include CVST. APS has been implicated in 6–17% of all cases of CVST and tends to predispose to CVST at a relatively younger age [44].

Vasculopathies, described in detail in Section 2, including Moyamoya and Sneddon's syndrome, overlap with APS at a rate of 21% and 50% respectively. Reversible cerebral vasoconstriction syndrome (RCVS) has also been described in patients with APS [46].

Other neurologic manifestations of the antiphospholipid syndrome include headache (20%), seizures (8%), and chorea (1.3–4.5%), with less frequent neurological manifestations including parkinsonism (especially progressive supranuclear palsy), dystonia, ballismus, myoclonus, cerebella ataxia, transverse myelitis, cognitive impairments, psychiatric symptoms, and peripheral neuropathy [4, 57].

3.4 Does the pattern of antibody positivity influence the likelihood of stroke?

As outlined in **Table 7**, some aPL are associated with a higher risk of ischemic stroke than others. Isolated LA positivity induces the greatest individual antibody risk for ischemic stroke [58]. Anti- β2-GPI were also associated with increased risk but to a lesser degree [58]. aCL and antiprothrombin antibodies have been reported variably with some studies showing no increased risk as an independent risk factor [27] while others reported to be independent risk when considering young patients exclusively [58]. As mentioned, triple positivity with positive LA, β2-GPI antibodies and aCL antibodies confers the highest risk [58].

High risk	Moderate risk	Low risk
Triple positivity (LA + aCL + anti-β2-GPI)	Isolated aCL when persistently positive in patients with SLE	Isolated anti-β2-GPI positivity
Isolated LA positivity		Inconsistent and low titer isolated aCL positivity

Table 7.
Risk for cerebrovascular event based on serologic profile. Adapted [58].

3.5 Does the presence of other risk factors for cerebrovascular events increase the risk in patients with APS?

Traditional cardiovascular risk factors also play a role in outcomes for patients with APS. Studies reveal that hypertension and smoking are the risk factors most associated with repeat thrombotic arterial events [59]. Combinations of risk factors have also been shown to increase the risk of repeat events [60]. Prospective studies evaluating the results of risk factor control have yet to be reported.

The RATIO study (Risk of Arterial Thrombosis In relation to Oral contraceptives) identified that the use of oral contraceptives (OCPs) and smoking carried an extremely high risk for women with APS in terms of risk for myocardial infarction and ischemic stroke [28]. The data revealed that the relative risk for ischemic stroke was higher in those who were smoking and in women with OCPs. The odds ratio for ischemic stroke was 43.1 (95%CI 12.2–152.0), which increased to 201.0 (95%CI 22.1–1828.0) in women who used oral contraceptives and 87.0 (14.5–523.0) in those who smoked. In women who had anti- β2-GPI, the risk of ischemic stroke was 2.3 (95%CI 1.4–3.7), but the risk of myocardial infarction was not increased (OR 0.9, 95%CI 0.5–1.6). Neither aCL nor anti-prothrombin antibodies affected the risk of myocardial infarction or ischemic stroke [28].

4. Etiology and mechanisms of stroke in APS

4.1 Pathophysiology of stroke in APS

Vascular thrombosis in APS can affect a wide variety of organ systems, but cerebrovascular thrombosis leading to stroke and transient ischemic attack is the most prevalent and perhaps the most consequential arterial event [61]. In a retrospective study of 135 APS patients, the highest morbidity was linked to neurologic involvement especially due to arterial thrombosis [62]. APS is also an important cause of stroke in the young, but as described can also affect older individuals [60]. The mechanisms of stroke in APS are diverse and include thrombosis in arteries, veins, and the microvasculature, as well as cardioembolism from non-bacterial thrombotic endocarditis.

The pathophysiology of vascular thrombosis in APS is not completely understood, but several studies suggest multiple converging pathways involving not only antibodies but also endothelial cells, platelets, monocytes, coagulation cascade proteins, and complements [63] producing a systemic thrombo-inflammatory state. The presence of aPL is not the sole cause for the significant clinical manifestations of APS as there can be asymptomatic "carriers" [17, 60]. Therefore, as previously mentioned, a "two-hit" hypothesis has been theorized, where the first-hit involves the presence of circulating aPL and associated endothelial dysfunction, and the second-hit presents an inflammatory insult such as trauma, surgery, or infection, leading to upregulation of β2GPI receptors on endothelial cells, as schematically demonstrated in **Figure 2**.

Even though aPL can be detected either by clotting tests, such as LA, or by an ELISA, such as aCL and anti-β2GPI, they are predominantly directed against β2GPI [17] and prothrombin [64]. Other important antigens recognized by aPL are annexin V, phosphatidylethanolamine, and phosphatidylserine [65]. Mechanistically these autoantibodies target phospholipid-binding plasma proteins bound to the surface of vascular endothelial cells and thrombocytes [60]. Plasma proteins predominantly bind to phosphatidylserine [17]. Normally located in the inner surface of cell membranes, phosphatidylserine becomes externalized when endothelial cells, platelets, and monocytes are activated. The avidity with which β2GPI binds to

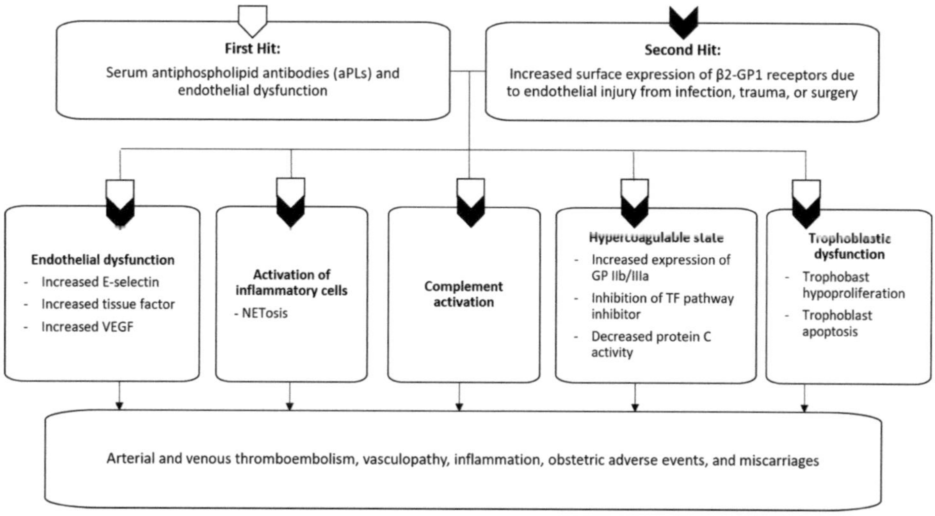

Figure 2.
The pathophysiology of vascular thrombosis in APS is not completely understood, but a 2-hit hypothesis is widely proposed. The first hit involves the presence of circulating aPL and endothelial injury, while the second hit requires an inflammatory insult such as trauma, surgery, or infection, leading to upregulation of beta-2 glycoprotein 1 (β2-GP1) receptors on endothelial cells. The aPLs-β2-GP1 receptor interaction unleashes multiple converging downstream pathways culminating in a thrombo-inflammatory state. VEGF: vascular endothelial growth factor; neutrophil extracellular traps (NETosis); GP: glycoprotein; TF: tissue factor (adapted [64, 66, 67]).

phosphatidylserine is further enhanced by the 'β2GPI'- 'β2GPI antibody dimerization' [66]. The downstream effect of β2GPI antibodies on endothelial cells and monocytes includes increased expression of tissue factor and thromboxane A2 which trigger the extrinsic coagulation pathway [64, 67]. Furthermore, the antibody binding inhibits the tissue factor pathway inhibitor and protein C activity [64, 67]. Taken together, the net effect is the synergistic production of a prothrombotic state. Endothelial cells, upon stimulation with aPL, also downregulate their nitric oxide production and increase the surface expression of adhesion molecules such as E-selectin leading to pro-inflammatory and pro-coagulation endothelial phenotype [17, 57, 67, 68]. This antibody-induced endothelial injury can lead to intimal hyperplasia, micro-vasculopathy, and accelerated atherosclerosis [69]. Activated platelets increase their surface expression of GPIIb-IIIa, synthesis of thromboxane A2 and platelet factor-4a, all acting to facilitate thrombosis [67]. Activation of neutrophils with accompanying release of Neutrophil Extracellular Traps (NETosis) and IL-8 may also play a role [67]. Annexin V, a natural anticoagulant, binds to phosphatidylserine (a procoagulant) forming an anticoagulant shield in the physiologic state in APS, this shield is disrupted tipping the system in favor of coagulation [70]. Upregulation in the mTOR (mechanistic target of rapamycin) pathway on endothelial cells may partly explain the microvascular thrombosis seen in APS.

In addition to vascular thrombosis, up to one-third of patients with APS develop non-bacterial thrombotic endocarditis (NBTE) in which there is a deposition of sterile platelet thrombi on heart valves, particularly the mitral and aortic valves, which can be a source of cardioembolic strokes [66].

4.2 Genetic considerations

Population and family studies, as well as animal studies, have suggested genetic disposition may be relevant to the development of APS. Like many autoimmune disorders, predisposition to APS has been mapped to genes in the major histocompatibility complex (MHC), among others. Also, epigenetic phenomena such as altered microRNA biogenesis in neutrophils, leading to accelerated atherosclerosis, have been implicated in APS [63].

5. Diagnostic workup for APS

The initial workup for stroke in the setting of APS is consistent with that of other stroke etiologies. Specifically, a multisystem approach evaluating from "*heart to head*" should be performed. However, in the setting of APS, a "*head to toe*" examination may be more aptly described. Prior to initiating an APS workup, there need to be history and examination findings that begin to clue the diagnostician towards an underlying process related to APS. Such findings, as previously mentioned in Section 2 and to be discussed, are important to consider before initiating an extensive and potentially costly workup. Although, among appropriate patients, APS should be considered in numerous stroke/cerebrovascular settings including acute ischemic infarct, hemorrhagic infarct, cerebral venous sinus thrombosis, and TIA.

5.1 When to test?

What raises the suspicion for APS in stroke? When should it be considered that more information and studies are needed besides the typical workup usually undertaken? The most pertinent situation would be when a younger patient (<50 years) presented with a thrombotic stroke without identified classic risk factors for ischemic/embolic stroke [71]. Initial workup may reveal exam and laboratory findings that may raise the concern for APS as listed in **Table 8**. Notably, subtle renal, cardiac, hematologic, and dermatologic system alterations can be indicative. Further, a family history of early-onset stroke, clotting, or other systemic features should be queried. Absence of typical risk factors including hypertension, diabetes, atrial fibrillation, or known history of coagulopathy (e.g. protein

1. Hematologic a. Thrombocytopenia i. Mild/common: platelets 50,000–100,000 cells per mm [3] ii. Severe/uncommon: platelets <20,000 cells per mm [3] b. Hemolytic Anemia i. Autoimmune hemolytic anemia (no schistocytes) ii. Thrombotic microangiopathy (with schistocytes)
2. Neurologic a. Cognitive impairment (with no evidence of stroke) b. Subcortical white matter change
3. Dermatologic a. Livedo reticularis or racemosa (consider Sneddon syndrome) b. Livedo vasculitis (painful, recurrent ulcerations of bilateral lower extremities
4. Cardiac a. Thickening (>3 mm) of the cardiac valves (proximal/middle part of valve leaflet, nodules with irregularity on atrial side of mitral valve or vascular side of aortic valve) b. Valve vegetations
5. Renal a. Acute kidney injury due to/or evidence of acute microangiopathic thrombosis

Table 8.
Other important clinical signs of APS not noted in Sapporo criteria, by body system. Adapted [63].

C deficiency, protein S deficiency, antithrombin III), among others, further increases the consideration for APS. Notably, as many as 17% of cardiovascular events in those under 50 reveal aPL antibodies and up to 22% including anticardiolipin antibodies [54].

Of note, without suggestion of underlying coagulopathy or clinical findings (see **Table 8**) a young patient without classic risk factors, testing for many coagulopathies is not routinely performed. When performed, there is also the question of whether this workup needs to occur in the inpatient setting, during the patient's admission for stroke, or if it can be done post-discharge. When considering this, the most important question is: Will any findings acutely change management? It should also be noted that for a positive diagnosis APS testing needs to occur multiple times over a 3 month or longer time period. If considering the APS diagnosis, formal hematology and/or rheumatology consult is recommended. In general, the recommendation for inpatient vs. outpatient is that some workup may be deferred if necessary, to the outpatient setting, either under the care of the patient's primary physician/provider, neurologist, hematologist, or rheumatologist.

5.2 What to test?

Consistent with all stroke patients, every patient should receive standard stroke workup testing including brain imaging (CT brain, MRI brain), vessel imaging of the head, neck, and great vessels of the chest (CTA, MRA), cardiac imaging including a transthoracic echocardiogram (TTE) and laboratory testing (CMP, CBC, PT/INR, aPTT, TSH, HgbA1C, lipid profile). A bubble study with the TTE should be considered if a paradoxical embolus from a DVT is on the differential. It is also recommended to obtain basic inflammatory markers such as sedimentation rate (ESR) and C-reactive protein (CRP) to evaluate for suggestion of diffuse inflammatory disease [24].

Transesophageal echocardiogram (TEE) should also be considered if the etiology remains uncertain, this is due to the increased frequency of valvular abnormalities in the setting of APS that may include irregular nodules/vegetations most commonly on the atrial side of the mitral valve or vascular side of the aortic valve, or if thickening of the valves is noted on TTE. Most commonly, the left side of the heart is the affected side with the mitral valve more commonly affected compared to the aortic valve. These cardiac changes are postulated to be due to immune complex damage and fibrosis [72].

If APS is being considered, it is recommended that while inpatient with the acute stroke the patient should have all antiphospholipid antibodies checked, according to the revised Sapporo laboratory criteria (see **Table 1**). Notably, this includes ELISA IgM/IgG for anticardiolipin (aCL) with a positive test showing medium to high titers (>40 GPL/MPL units or >99th percentile), which will need to be confirmed on at least two or more occasions, 12-weeks apart. Lupus anticoagulant (LA) should also be checked by two tests including dilute Russell viper venom time (dRVVT) and LA-sensitive PTT (PTT-LA)), again conformed on at least two occasions, 12-weeks apart. Lastly, an ELISA IgM/IgG for anti-beta2-glycoprotein I (β2GPI) should also be tested, with a positive value determined by titer in the 99th percentile, and again, should be tested on at least two occasions 12-weeks apart.

At least one clinical criterion (in the context of this chapter, most likely stroke) and one laboratory criterion should be met to diagnosis APS. As described, these tests are done 12-weeks apart, so the first set of lab tests will be performed inpatient and then the second 12-weeks later, typically performed in the outpatient setting.

As outlined in **Table 8**, if the patient does not meet revised Sapporo criteria, APS may still be diagnosed if clinical suspicion remains high based on multi-system abnormalities and if further etiologies are not identified [64].

If a patient inconsistently tests positive for APS, it may be warranted to also check for other autoimmune diseases, namely systemic lupus erythematosus (SLE), as up to 36% of those with APS will be positive for SLE. Having both APS and SLE increases the risk for stroke beyond having only one or the other [31].

5.3 Understanding the tests

As described above, there are 3 primary antibody tests for APS including aCL, LA, and β2GPI. Anticardiolipin (aCL) testing was first developed as a test for syphilis in the 1900s [71]. The aCL antibody was found not to be specific to just syphilis, thus its utility as a test for APS was also found after many false-positive syphilis tests showed an increased risk for thrombotic events. The tests presently use tissue derived from bovine tissue. Both IgG and IgM are evaluated by ELISA for the presence of aCL antibodies. Notably, due to cross-reactivity as discussed with syphilis, the presence of aCL does not alone confirm APS.

Lupus anticoagulant (LA) is a test for immunoglobulins that while associated with thrombosis, are associated with preventing coagulation in vivo. The process for testing LA is three tests including screening (usually with aPTT or dRVVT, clotting of phospholipid factors), mixing (correct with normal plasma), and confirmation (shortening prolongation with added phospholipid) [67]. Once again, LA by itself cannot confirm APS due to cross-reactivity. LA testing is outlined in **Figure 3**.

Anti-β2 glycoprotein I (β2GPI) enzyme-linked immunosorbent assay (ELISA) testing is the last of the trio of tests for APS. There are 5 main domains of the β2GPI,

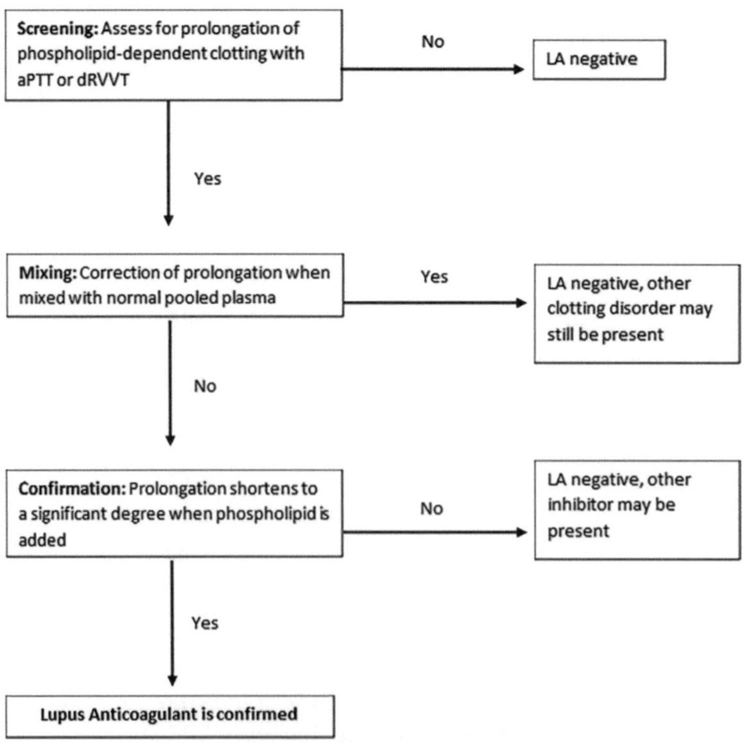

Figure 3.
Testing for lupus anticoagulant (Adapted [67]).

labeled DI through DV. Anti-β2GPI largely targets domain I (DI). When this domain is targeted, it has been shown an association with thrombosis. The other domains DII through DV being targeted have not been shown to have as strong a connection for promoting thrombosis. Of note, there are some more rare entities that may also raise anti-β2GPI levels, such as leishmaniasis, leptospirosis, or leprosy. For APS, the associated antibodies are against the IgG form, whereas other elevates of anti-β2GPI may be directed towards the IgM variety [73].

5.4 Implications of acute diagnosis

Unless the patient presents with a prior history of APS, the diagnosis of APS will likely be in question during the acute and subacute stroke window. This is because APS by laboratory criteria needs to be performed 12-weeks apart with two positive tests to confirm. That said, a patient that presents with a stroke and has one or more laboratory results that are concerning for APS (positive LA, aCL, anti-β2GPI), there is a question if confirming APS would change acute management. Oftentimes, the answer is yes; this even in the setting of likely APS, because thrombosis can be multifactorial and can progress between confirmatory APS testing [67]. As such, management should focus on appropriate treatment for the source of the stroke. For example, if the source is cardioembolic, the timing of initiation of anticoagulation should be considered, weighing the risk of a second embolic event while not on indicated therapy versus the risk of hemorrhagic conversion of the primary infarct.

6. Treatment: primary and secondary prevention

Once the workup for APS is complete, and if positive, the next logical step is to address treatment. However, prior to addressing treatment, let us first consider if APS is a primary risk factor for stroke risk. Numerous studies have been performed to address this question, culminating with a meta-analysis evaluating 15 different studies in aggregate [54]. In this evaluation, 13 of the 15 studies reported a significant association between a CVE and aPL antibodies with a cumulative odds ratio of 5.48 [54]. While this study provides insight into primary event risk, a follow-up question relates to the risk of APS with recurrent stroke. A second meta-analysis was completed looking at 8 studies to answer this question, demonstrating no statistically significant risk of recurrent ischemic stroke among APS patients [74]. Understanding why one meta-analysis demonstrated a link between aPL antibodies and single ischemic events, while another did not show a link with recurrent events remains challenging to understand. One hypothesis used to explain these incongruent findings is that clinical events do not occur frequently occur despite the presence of the antibodies, suggesting that treatment and/or lifestyle modifications after a first stroke affect the chance of a second event [74, 75]. Therefore, an understanding that APS is associated with the single cerebral vascular event, and that treatment affects the chance of a second event, indicates that secondary prevention is highly warranted.

6.1 Primary prevention

Knowing that therapy is indicated, we can now evaluate various treatments on the risk of thrombosis in the setting of APS. In those individuals without any other risk factors, the risk of thrombosis is less than 1% per year [76, 77]. In this group, when they do present with a thrombus, it is normally in the setting of another thrombotic risk factor, such as cancer, surgery, pregnancy, estrogen use, acute infection, smoking, and hypertension. On the other hand, the risk of thrombosis can be as high as 5% per year in individuals with a persistent moderate high-risk profile including aPL antibodies and a systemic autoimmune disease [78]. Therefore, with the risk of thrombosis being so variable, sometimes as low as 1% or other times as high as 5%, the question of optimal prevention strategies can be challenging.

Regarding primary prevention (before a stroke or vascular event) the answer remains controversial with only scant data based on prospective trials [79]. Some of these trials have demonstrated a decrease in thrombosis with the use of aspirin. For example, a meta-analysis of 11 mostly observational studies demonstrated a 2-fold risk reduction in the first thrombotic event with a more significant effect in those with arterial thrombosis [79]. Post subgroup analysis of only prospective trials demonstrated there was no significant difference between aspirin and those not treated [79]. Therefore, with conflicting data on aspirin, one may ask could there be a benefit with the use of anticoagulation as well as aspirin for primary prevention. While the data was limited, one primary prevention study evaluated the use of aspirin alone vs. aspirin plus anticoagulation in 166 patients, demonstrating no significant difference in terms of the amount of thrombotic events between groups, with an increased risk of bleeding in the aspirin plus warfarin arm [80]. Therefore, given the increased bleeding risk, the use of aspirin and warfarin in combination is not recommended for primary prevention, with the question of aspirin use in isolation remaining. Many agencies have weighed in on this subject including the 13th International Congress on Antiphospholipid Antibodies as well as the European League Against Rheumatism making recommendations suggesting the use of aspirin in high-risk antiphospholipid profiles, those with other thrombotic risk factors, as well as those with SLE [58, 81]. Even with these recommendations, one must also consider the risk of bleeding with the use of aspirin. One meta-analysis looking at six randomized control trials showed an association of increased annual risk of major bleeding in those patients using aspirin with hypertension, age > 65, diabetes, and male sex being the most significant associated risk factors [82].

In summary, the decision to use primary prevention remains an individualized choice based on a patient-centric decision. Overall, though one should consider the use of primary prevention with aspirin in those with cardiac risk factors, high risk antiphospholipid antibody profile, presence of other thrombotic risk factors and in the presence of other autoimmune disease always ensuring a thorough risk benefit analysis is done with concern for bleeding. See **Figure 4** for breakdown of treatment option algorithm.

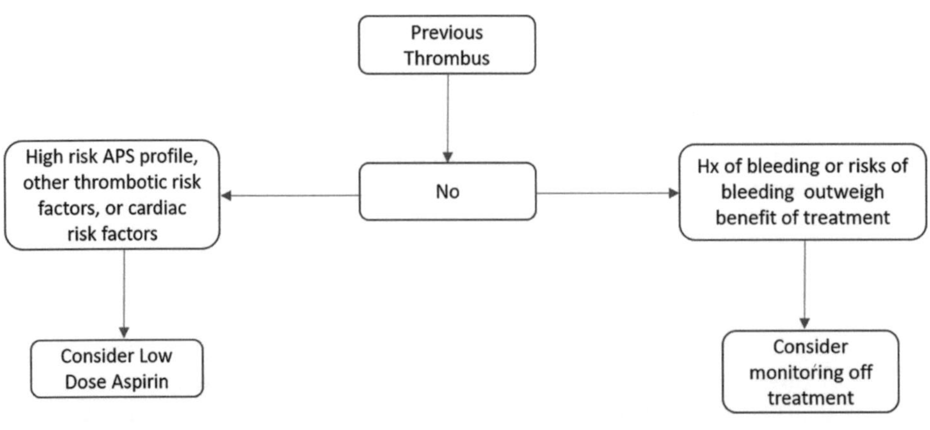

Figure 4.
Treatment options algorithm (adapted [10]).

6.2 Secondary prevention: arterial primary event

Knowing the indications for the use of primary prophylaxis we now consider secondary prophylaxis. Data regarding the need for secondary prophylaxis specifically in previous arterial thrombi remains scant without any consensus. For example, one study demonstrated the use of warfarin with a goal INR of 1.4–2.8 was not superior to full dose aspirin 325 mg alone for stroke prevention, with concerns that this study was flawed due to transient positivity of aPL antibodies [27]. Another study evaluating 20 patients with ischemic stroke demonstrated that the use of low-dose aspirin and warfarin with a goal INR of 2–3 was superior to low-dose aspirin alone in the prevention of further arterial thrombi [11]. While two other studies demonstrated that for older patients with stroke, and a single test showing low titers of anticardiolipin antibodies, that aspirin may be as effective as warfarin [27, 83]. With this conflicting data, there remains no consensus statement on secondary prophylaxis with many agencies weighing in on this subject. For example, the 13th International Congress on Antiphospholipid Antibodies as well as the European League Against Rheumatism both recommended secondary prophylaxis with high-intensity warfarin with an INR > 3 or low dose aspirin combined with moderate-intensity warfarin with an INR from 2 to 3 [58, 81]. Both agencies decided on using a goal INR of >3 for warfarin because in previous studies evaluating different doses of warfarin in treating thrombi, relatively few patients with arterial thrombi were enrolled [84, 85]. Overall, data remains scarce and guidelines are based upon a consensus of expert opinion. In those with recurrent arterial events, some recommend increasing target INR level and or switching to low molecular weight heparin with the addition of other adjective therapies to include statins [86].

In summary, the decision on which patient to treat and which agent to use for secondary prophylaxis with arterial thrombi remains a patient-centric decision. Those with high-risk aPL profiles, presence of other systemic autoimmune diseases, and or other risk factors for thrombus would likely benefit from treatment with either aspirin and warfarin with a goal 2–3 or warfarin alone with a goal INR 3–4. Those with recurrent events would likely benefit from increasing the INR goal or if not feasible switching to low molecular weight heparin. Moving forward it would be beneficial to validate a risk stratification model to identify those with arterial thrombosis who would benefit from more aggressive treatment [67]. See **Figure 5** demonstrates a treatment options algorithm.

6.3 Secondary prevention: venous primary event

Now knowing the indications and treatment options for the use in secondary arterial prophylaxis we now move on to secondary venous prophylaxis, which in the case of stroke would be beneficial in treating paradoxical emboli. Much different from that in arterial secondary prophylaxis, there is more of a consensus regarding the treatment of secondary venous prophylaxis using warfarin with a goal INR of 2–3 showing a decrease in recurrent venous events of 80–90% [57, 87]. Some studies have evaluated the use of higher intensity anticoagulation with a goal INR of 3.1–4.5 showing no reduced risk in thrombosis, but a significant excess of minor bleeding [84, 85].

Therefore, with the above data, we can safely say in summary for secondary prevention for venous thrombi in those with a chance of paradoxical emboli treatment with warfarin with a goal INR of 2–3 is indicated. See **Figure 5** for a treatment options algorithm.

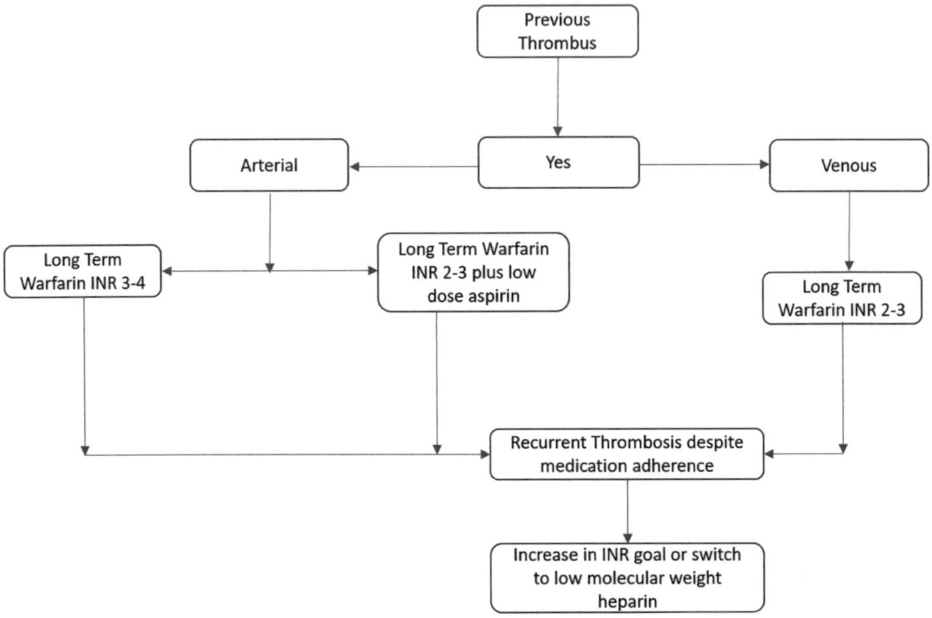

Figure 5.
Arterial versus venous thrombus treatment options algorithm (adapted [13]).

6.4 Other treatment considerations

6.4.1 Direct oral anticoagulants

Following the basics of both primary and secondary prevention, one may question other anticoagulation options as adjuvant therapies. Regarding the use of direct oral anticoagulants (DOACs) there remains insufficient evidence with data suggesting an increased risk of thrombosis [88]. For example, two studies demonstrated no difference in the rate of venous thromboembolism and an increased risk of arterial thrombotic with the use of rivaroxaban over warfarin [89, 90]. Looking at this data more closely, a meta-analysis of these two studies did not find an increased risk of thrombosis in patients treated with rivaroxaban over warfarin at a 6 month follow up, however for unclear reasons, almost 3/4 of the thrombi occurred post the 6 months follow up [39]. Given the lack of prospective data, the utility of DOACs in the treatment of thrombus formation remains uncertain.

6.4.2 Other therapies

Beyond DOACs, other adjuvant therapies have been studied including statins and hydroxychloroquine. With statins being a mainstay of treatment post-stroke, it would not be unreasonable to think that they may be beneficial in APS, potentially exhibiting pleiotropic effects including anti-inflammatory, antithrombotic, and as well as the expected lipid-lowering potential [13]. To date, there have been no randomized controlled trials looking at the efficacy in this group of patients. One study however did look at the levels of pro-inflammatory and prothrombotic markers post use of Fluvastatin, which were significantly decreased suggesting their benefit in APS [91]. At this time without a randomized control trial, the 15th International Congress on Antiphospholipid Antibodies has recommended the use of statins in those with high cardiovascular risks and or recurrent thrombosis despite adequate AC [88]. Regarding the use of hydroxychloroquine, similar to statins in addition to its immunomodulatory effect, it also has antithrombotic properties making it a good candidate as adjunctive therapy [88]. Two studies have been performed demonstrating differing results regarding treatment with hydroxychloroquine plus aspirin vs. aspirin alone. The first demonstrated no difference between rates of thrombosis between both groups [92]. The other demonstrated a significantly lower thrombotic rate compared to standard of care alone, in addition to down-trending antibody titers [93]. These data suggest that both statins and hydroxychloroquine could be beneficial as adjunctive therapies in specific situations, although more data is needed for consensus.

6.4.3 Stopping therapy

Throughout this section, we have addressed the need for primary and secondary prevention, but one question left unanswered is safety as associated with therapy cessation. Unfortunately, there remains a multitude of answers to this question, hence each case should be considered independently. In those with a history of arterial thrombotic events, the risk of repeat thrombus formation off anticoagulation is too high and therefore indefinite anticoagulation is warranted [94]. In those with a history of transient positivity of antiphospholipid antibodies who eventually become negative based on two separate studies, one can consider stopping anticoagulation [95, 96]. Specifically, this would be associated with those who only have primary APS with persistently negative antibodies where if there was a thrombotic event it occurred in association with a transient risk factor including pregnancy or immobilization as examples [96]. In these cases, it is thought that the antibodies do not play a pathogenic role, but rather are a "phenomenon". Therefore, some have recommended a 3–6-month course of anticoagulation with consideration to look for residual thrombus, which has been shown to increase the rate of recurrence by 50% [94]. Notably, the data and recommendations regarding stopping anticoagulation are based upon two small case series. Therefore, with such insufficient data, unless the risk of anticoagulation outweighs the benefit it would not be recommended to stop anticoagulation in those that become persistently negative.

6.4.4 Final thoughts on therapy

Throughout this section we have addressed both preventions of stroke in APS, but what if someone should fail prevention and come in with an acute stroke. The answer to this question unlike many of the other is simple. Acute management is no different than those with or without APS [97]. Lastly, as described, APS often requires treatment with anticoagulant medications such as heparin to reduce the risk of further episodes of thrombosis and improve the prognosis of pregnancy. Warfarin (brand name Coumadin) should not be used during pregnancy because it crosses the placenta and is teratogenic. Unfractionated heparin (UFH) and low molecular weight heparin do not cross the placenta and are safe for the fetus, but long-term treatment with UFH is problematic because of its inconvenient administration, the need to monitor anticoagulant activity, and because of its potential side effects, such as heparin-induced thrombocytopenia and osteoporosis [98].

7. Conclusion

Thromboses of the cerebral arterial and venous systems are a common manifestation of APS leading to ischemic and/or hemorrhagic stroke. APS has been a recognized cause of CVE especially in those without classic cardiovascular risk factors. It has been estimated that one in five strokes and patients younger than 45 could be associated with APS and some newer studies show that APL antibodies are present in approximately 14% of stroke patients. Persistently elevated APL seems to increase the risk for CV by at least fourfold. Stroke is the fourth most common presenting symptom behind deep venous thrombosis, thrombocytopenia, and livedo reticularis. The recurrent risk of stroke in APS patients has been less widely studied as compared to other types of thromboses, however, cumulative risk of 14% for brain ischemia at 10 years has been reported. APS increases stroke risk via many mechanisms including hypercoagulability, inflammation, accelerated atherosclerosis, and cardiac manifestations, among others. Mechanistically these lead to in-situ clot formation and/or embolic phenomena. Physicians must carefully consider all these potential mechanisms when evaluating and treating stroke patients to achieve both optimal short- and long-term outcomes. While the exact underlying pathophysiology of APS remains uncertain, underlying genetics in the setting of a triggering event (e.g., surgery, trauma, infection) is believed to play a key role in the development of the disease. While primary and secondary prevention recommendations continue to evolve, each case should be considered independently to achieve optimal results. Results from more randomized control trials are needed to further infer upon the ever-evolving consensus guidelines. For the time being, the decision to use primary and/or secondary prevention therapies, and of which type, will continue to be an individualized patient-centric decision requiring careful interpretation of test results with multispecialty (neurology, hematology, rheumatology) input.

Author details

Kathryn Grimes[1,2], Adam P. Klein[1,2], Rakhee Lalla[1,2], Adeolu Morawo[1,2], Sana Somani[1,2], Matthew J. Woodward[1,2] and John W. Cole[1,2]*

1 University of Maryland School of Medicine, Baltimore, MD, USA

2 Baltimore Veterans Affairs Medical Center, Baltimore, MD, USA

*Address all correspondence to: jcole@som.umaryland.edu

IntechOpen

© 2022 The Author(s). Licensee IntechOpen. This chapter is distributed under the terms of the Creative Commons Attribution License (http://creativecommons.org/licenses/by/3.0), which permits unrestricted use, distribution, and reproduction in any medium, provided the original work is properly cited.

References

[1] Miyakis S, Lockshin MD, Atsumi T, Branch DW, Brey RL, Cervera R, et al. International consensus statement on an update of the classification criteria for definite antiphospholipid syndrome (APS). Journal of Thrombosis and Haemostasis. 2006;**4**:295-306. DOI: 10.1111/j.1538-7836.2006.01753.x

[2] Žigon P. Do antiphospholipid antibodies enhance thromboembolic risk in patients with cancer? Polish Archives of Internal Medicine. 2020; **130**(12):1026-1028. DOI: 10.20452/pamw.15724. Epub 2020 Dec 22

[3] Khangura RK, Cooper S, Luo GY. Antiphospholipid antibody syndrome: Pathogenesis, diagnosis, and management in pregnancy. Maternal-Fetal Medicine. 2019;**1**:38-42. DOI: 10.1097/FM9.0000000000000007

[4] Lim W. Antiphospholipid syndrome. Hematology. American Society of Hematology. Education Program. 2013; **1**:675-680. DOI: 10.1182/asheducation-2013.1.675

[5] Fleetwood T, Cantello R, Comi C. Antiphospholipid syndrome and the neurologist: From pathogenesis to therapy. Frontiers in Neurology. 2018;**9**: 1001. DOI: 10.3389/fneur.2018.01001

[6] Misita CP, Moll S. Antiphospholipid antibodies. Circulation. 2005;**112**:39-44. DOI: 10.1161/CIRCULATIONAHA.105.548495

[7] Sciascia S, Radin M, Cecchi I, Rubini E, Scotta A, Rolla R, et al. Reliability of lupus anticoagulant and anti-phosphatidylserine/prothrombin autoantibodies in antiphospholipid syndrome: A multicenter study. Frontiers in Immunology. 2019;**10**:376. DOI: 10.3389/fimmu.2019.00376

[8] De Laat B, Pengo V, Pabinger I, Musial J, Voskuyl AE, Bultink IE, et al. The association between circulating antibodies against domain I of beta2-glycoprotein I and thrombosis: an international multicenter study. Journal of Thrombosis and Haemostasis. Nov 2009;7(11):1767-1773

[9] Wilson WA, Gharavi AE, Koike T, et al. International consensus statement on preliminary classification criteria for definite antiphospholipid syndrome: Report of an international workshop. Arthritis and Rheumatism. 1999;**42**(7): 1309-1311. DOI: 10.1002/1529-0131(199907)42:7<1309::AID-ANR1>3.0.CO;2-F

[10] Ghembaza A, Saadoun D. Management of antiphospholipid syndrome. Biomedicines. 2020;**8**(11):508

[11] Okuma H, Kitagawa Y, Yasuda T, Tokuoka K, Takagi S. Comparison between single antiplatelet therapy and combination of antiplatelet and anticoagulation therapy for secondary prevention in ischemic stroke patients with antiphospholipid syndrome. International Journal of Medical Sciences. 2010;**7**(1):15

[12] Uthman I, Noureldine MHA, Ruiz-Irastorza G, Khamashta M. Management of antiphospholipid syndrome. Annals of the Rheumatic Diseases. 2019;**78**(2): 155-161. DOI: 10.1136/annrheumdis-2018-213846. Epub 2018 Oct 3

[13] Khamashta M, Taraborelli S, Sciascia A, Tincani. Antiphospholipid Syndrome. Best Practice & Research Clinical Rheumatology. 2016;**30**: 133-148. DOI: 10.1016/j.berh.2016.04.002

[14] Cervera R, Serrano R, Pons-Estel GJ, Ceberio-Hualde L, Shoenfeld Y, de Ramón E, et al. Morbidity and mortality in the antiphospholipid syndrome during a 10-year period: A multicentre prospective study of 1000 patients.

Annals of the Rheumatic Diseases. 2015; **74**(6):1011-1018. DOI: 10.1136/annrheumdis-2013-2048

[15] Kaul M, Erkan D, Sammaritano L, Lockshin MD. Assessment of the 2006 revised antiphospholipid syndrome classification criteria. Annals of the Rheumatic Diseases. 1 Jul 2007;**66**(7): 927-930

[16] Caplan L, Biller J. Uncommon Causes of Stroke. 3rd ed. Cambridge, United Kingdom: Cambridge University Press; 2018. pp. 305-316

[17] Cervera R. Antiphospholipid syndrome. Thrombosis Research. 2017; **151**(Suppl 1):S43-s47

[18] Aguiar CL, Erkan D. Catastrophic antiphospholipid syndrome: How to diagnose a rare but highly fatal disease. Therapeutic Advances in Musculoskeletal Disease. 2013;**5**(6):305-314. DOI: 10.1177/1759720X13502919

[19] Chinnery PF, Shaw PJ, Ince PG, et al. Fulminant encephalopathy due to the catastrophic primary antiphospholipid syndrome. Journal of Neurology, Neurosurgery, and Psychiatry. 1997;**62**(3):300-301. DOI: 10.1136/jnnp.62.3.300

[20] Arachchillage DRJ, Laffan M. Pathogenesis and management of antiphospholipid syndrome. British Journal of Haematology. 2017;**178**(2): 181-195. DOI: 10.1111/bjh.14632. Epub 2017 Mar 24

[21] Gómez-Puerta JA, Cervera R. Diagnosis and classification of the antiphospholipid syndrome. Journal of Autoimmunity. 2014;**48-49**:20-25. DOI: 10.1016/j.jaut.2014.01.006. Epub 2014 Jan 24

[22] Andreoli L, Chighizola CB, Banzato A, et al. Estimated frequency of antiphospholipid antibodies in patients with pregnancy morbidity, stroke, myocardial infarction, and deep vein thrombosis: A critical review of the literature. Arthritis Care & Research. 2013;**65**(11):1869-1873. DOI: 10.1002/acr.22066

[23] Hughes G. Migraine, memory loss, and "multiple sclerosis"; Neurological features of the antiphospholipid (Hughes') syndrome. Postgraduate Medical Journal. 2003;**79**(928):81-83

[24] Panichpisal K, Rozner E, Levine SR. The management of stroke in antiphospholipid syndrome. Current Rheumatology Reports. 2012;**14**:99-106

[25] Janardhan V, Wolf PA, Kase CS, et al. Anticardiolipin antibodies and risk of ischemic stroke and transient ischemic attack: The Framingham cohort and offspring study. Stroke. 2004;**35**(3):736-741. DOI: 10.1161/01.STR.0000117575.48205.2D. Epub 2004 Feb 5

[26] Cojocaru IM, Cojocaru M, Burcin C, et al. Evaluation of antiphospholipid antibodies in young women with ischemic stroke. Romanian Journal of Internal Medicine. 2007;**45**(2):201-204

[27] Levine SR, Brey RL, Tilley BC, et al. Antiphospholipid antibodies and subsequent thrombo-occlusive events in patients with ischemic stroke. Journal of the American Medical Association. 2004;**291**(5):576-584. DOI: 10.1001/jama.291.5.576

[28] Urbanus RT, Siegerink B, Roest M, Rosendaal FR, de Groot PG, Algra A. Antiphospholipid antibodies and risk of myocardial infarction and ischaemic stroke in young women in the RATIO study: A case-control study. The Lancet Neurology. 2009;**8**(11):998-1005. DOI: 10.1016/S1474-4422(09)70239-X

[29] de Souza AW, Silva NP, de Carvalho JF, et al. Impact of hypertension and hyperhomocystenemia on arterial thrombosis in primary antiphospholipid syndrome. Lupus. 2007;**16**:782-787

[30] Pezzini A, Grassi M, Lodigiani C, et al. Predictors of long-term recurrent vascular events after ischemic stroke at young age: The Italian Project on Stroke in Young Adults. Circulation. 2014;**129**(16):1668-1676. DOI: 10.1161/CIRCULATIONAHA.113.005663. Epub

[31] de Amorim LC, Maia FM, Rodrigues CE. Stroke in systemic lupus erythematosus and antiphospholipid syndrome: Risk factors, clinical manifestations, neuroimaging, and treatment. Lupus. 2017;**26**(5):529-536. DOI: 10.1177/0961203316688784

[32] Babikian VL. Clinical and laboratory findings in patients with antiphospholipid antibodies and cerebral ischemia. The Antiphospholipid Antibodies in Stroke Study Group. Stroke. 1990;**21**(9):1268-1273

[33] Provenzale JM, Barboriak DP, Allen NB, Ortel TL. Antiphospholipid antibodies: Findings at arteriography. American Journal of Neuroradiology. 1998;**19**:611-616

[34] Kitagawa Y. Antiphospholipid syndrome and stroke. Rinshō Shinkeigaku. 2005;**45**(11):852-855

[35] Saidi S, Mahjoub T, Almawi WY. Lupus anticoagulants and anti-phospholipid antibodies as risk factors for a first episode of ischemic stroke. Journal of Thrombosis and Haemostasis. 2009;7(7):1075-1080. DOI: 10.1111/j.1538-7836.2009.03446.x. Epub 2009 Apr 27

[36] Atsumi T, Ieko M, Bertolaccini ML, et al. Association of autoantibodies against the phosphatidylserine-prothrombin complex with manifestations of the antiphospholipid syndrome and with the presence of lupus anticoagulant. Arthritis and Rheumatism. 2000;**43**(9):1982-1993. DOI: 10.1002/1529-0131(200009)43:9 <1982::AID-ANR9>3.0.CO;2-2

[37] Lopez LR, Dier KJ, Lopez D, et al. Anti-beta 2-glycoprotein I and antiphosphatidylserine antibodies are predictors of arterial thrombosis in patients with antiphospholipid syndrome. American Journal of Clinical Pathology. 2004;**121**(1):142-149. DOI: 10.1309/YVQ6-PX76-XMYM-3J29

[38] Ruffatti A, Calligaro A, Hoxha A, et al. Laboratory and clinical features of pregnant women with antiphospholipid syndrome and neonatal outcome. Arthritis Care & Research. 2010;**62**: 302-307

[39] Sanchez-Redondo J, Espinosa G, Delgado DV, Cervera R. Recurrent thrombosis with direct oral anticoagulants in antiphospholipid syndrome: A systematic literature review and meta-analysis. Clinical Therapeutics. 2019;**41**(9):1839-1862

[40] Rosove MH, Brewer PM. Antiphospholipid thrombosis: Clinical course after the first thrombotic event in 70 patients. Annals of Internal Medicine. 1992;**117**(4):303-308. DOI: 10.7326/0003-4819-117-4-303

[41] Sciascia S, Sanna G, Murru V, et al. GAPSS: The global anti-phospholipid syndrome score. Rheumatology (Oxford, England). 2013;**52**(8): 1397-1403. DOI: 10.1093/rheumatology/kes388. Epub 2013 Jan 12

[42] Radin M, Sciascia S, Erkan D, et al. The adjusted global antiphospholipid syndrome score (aGAPSS) and the risk of recurrent thrombosis: Results from the APS ACTION cohort. Seminars in Arthritis and Rheumatism. 2019;**49**(3): 464-468. DOI: 10.1016/j.semarthrit.2019.04.009

[43] Martinelli I, Battaglioli T, Pedotti P, Cattaneo M, Mannucci PM. Hyperhomocysteinemia in cerebral vein thrombosis. Blood. 2003;**102**(4): 1363-1366

[44] Silvis SM, Sousa DA, De FJM, Coutinho JM. Cerebral venous thrombosis. Nature Reviews. Neurology. 2017;**13**(9):555-565

[45] Carhuapoma JR, Mitsias P, Levine SR. Cerebral venous thrombosis and anticardiolipin antibodies. Stroke. 1997; **28**(12):2363-2369. DOI: 10.1161/01.str.28.12.2363

[46] Gupta S, Zivadinov R, Ramasamy D, et al. Reversible cerebral vasoconstriction syndrome (RCVS) in antiphospholipid antibody syndrome (APLA): The role of centrally acting vasodilators. Case series and review of literature. Clinical Rheumatology. 2014; **33**(12):1829-1833

[47] Wang Z, Fu Z, Wang J, et al. Moyamoya syndrome with antiphospholipid antibodies: A case report and literature review. Lupus. 2014;**23**(11):1204-1206. DOI: 10.1177/0961203314540761. Epub 2014 Jun 17

[48] Samanta D, Cobb S, Arya K. Sneddon syndrome: A comprehensive overview. Journal of Stroke and Cerebrovascular Diseases. 2019;**28**(8): 2098-2108. DOI: 10.1016/j.jstrokecerebrovasdis.2019.05.013. Epub 2019 May 31

[49] Wu S, Xu Z, Liang H. Sneddon's syndrome: A comprehensive review of the literature. Orphanet Journal of Rare Diseases. 2014;**9**(1):768

[50] Yelnik CM, Kozora E, Appenzeller S. Non-stroke central neurologic manifestations in antiphospholipid syndrome. Current Rheumatology Reports. 2016;**18**(2):11

[51] Appenzeller S, Lapa AT, Guirau CR, et al. Cognitive impairment in antiphospholipid syndrome: Evidence from animal models. Clinical Rheumatology. 2012;**31**(3): 403-406

[52] Cervera R, Piette J-C, Font J, Khamashta MA, Shoenfeld Y, Camps MT, et al. Antiphospholipid syndrome: Clinical and immunologic manifestations and patterns of disease expression in a cohort of 1,000 patients. Arthritis and Rheumatism. 2002;**46**(4): 1019-1027

[53] Ricarte IF, Dutra LA, Abrantes FF, Toso FF, Barsottini OGP, Silva GS, et al. Neurologic manifestations of antiphospholipid syndrome. Lupus. 2018;**27**(9):1404-1414. DOI: 10.1177/0961203318776110

[54] Sciascia S, Sanna G, Khamashta MA, Cuadrado MJ, Erkan D, Andreoli L, et al. The estimated frequency of antiphospholipid antibodies in young adults with cerebrovascular events: A systematic review. Annals of the Rheumatic Diseases. 2015;**74**(11): 2028-2033. DOI: 10.1136/annrheumdis-2014-205663

[55] Gašperšič N, Zaletel M, Kobal J, Žigon P, Čučnik S, Šemrl SS, et al. Stroke and antiphospholipid syndrome-antiphospholipid antibodies are a risk factor for an ischemic cerebrovascular event. Clinical Rheumatology. 2019; **38**(2):379-384. DOI: 10.1007/s10067-018-4247-3. Epub 2018 Aug 7

[56] Mehta T, Hussain M, Sheth K, Ding Y, McCullough LD. Risk of hemorrhagic transformation after ischemic stroke in patients with antiphospholipid antibody syndrome. Neurological Research. 2017; **39**(6):477-483. DOI: 10.1080/01616412.2017.1323382

[57] Leal Rato M, Bandeira M, Romão VC, Aguiar de Sousa D. Neurologic manifestations of the antiphospholipid syndrome—An update. Current Neurology and Neuroscience Reports. 2021;**21**(8):41. DOI: 10.1007/s11910-021-01124-z

[58] Ruiz-Irastorza G, Cuadrado M, Ruiz-Arruza I, Brey R, Crowther M,

Derksen R, et al. Evidence-based recommendations for the prevention and long-term management of thrombosis in antiphospholipid antibody-positive patients: Report of a Task Force at the 13th International Congress on Antiphospholipid Antibodies. Lupus. 2011;**20**(2):206-218. DOI: 10.1177/0961203310395803

[59] da Silva FF, Levy RA, de Carvalho JF. Cardiovascular risk factors in the antiphospholipid syndrome. Journal of Immunology Research. 2014; **2014**:621270. DOI: 10.1155/2014/621270

[60] Erkan D. A cross-sectional study of clinical thrombotic risk factors and preventive treatments in antiphospholipid syndrome. Rheumatology. 2002;**41**(8):924-929. DOI: 10.1093/rheumatology/41.8.924

[61] Ruiz-Irastorza G et al. Antiphospholipid syndrome. Lancet. 2010;**376**(9751):1498-1509

[62] Grika EP et al. Morbidity, mortality, and organ damage in patients with antiphospholipid syndrome. The Journal of Rheumatology. 2012;**39**(3):516-523

[63] Negrini S et al. The antiphospholipid syndrome: From pathophysiology to treatment. Clinical and Experimental Medicine. 2017;**17**(3): 257-267

[64] Sciascia S, Sanna G, Murru V, Roccatello D, Khamashta MA, Bertolaccini ML. Anti-prothrombin (aPT) and anti-phosphatidylserine/ prothrombin (aPS/PT) antibodies and the risk of thrombosis in the antiphospholipid syndrome. A systematic review. Thrombosis and Haemostasis. 2014;**111**(2):354-364. DOI: 10.1160/TH13-06-0509. Epub 2013 Oct 31

[65] Iuliano A, Galeazzi M, Sebastiani GD. Antiphospholipid syndrome's genetic and epigenetic aspects. Autoimmunity Reviews. 2019;**18**(9): 102352

[66] Schreiber K et al. Antiphospholipid syndrome. Nature Reviews Disease Primers. 2018;**4**:17103

[67] Garcia D, Erkan D. Diagnosis and management of the antiphospholipid syndrome. The New England Journal of Medicine. 2018;**378**(21):2010-2021. DOI: 10.1056/NEJMra1705454

[68] de Groot PG, de Laat B. Mechanisms of thrombosis in systemic lupus erythematosus and antiphospholipid syndrome. Best Practice & Research. Clinical Rheumatology. 2017;**31**(3): 334-341

[69] Corban MT et al. Antiphospholipid syndrome: Role of vascular endothelial cells and implications for risk stratification and targeted therapeutics. Journal of the American College of Cardiology. 2017;**69**(18):2317-2330

[70] Irman Š, Škarabot M, Muševič I, Rozman B, Božič B. Thrombomodulatory effect of anti-B2-glycoprotein I antibodies on crystalline annexin A5 on phospholipid bilayers, as observed by atomic force microscopy. EJIFCC. 2011;**21**(4):81-93

[71] Pengo V, Tripodi A, Reber G, Rand JH, Ortel TL, Galli M, et al. Update of the guidelines for lupus anticoagulant detection. Subcommittee on Lupus Anticoagulant/Antiphospholipid Antibody of the Scientific and Standardisation Committee of the International Society on Thrombosis and Haemostasis. Journal of Thrombosis and Haemostasis. 2009;**7**(10):1737-1740. DOI: 10.1111/j.1538-7836.2009.03555.x. Epub 2009 Jul 17

[72] Levine JS, Branch DW, Rauch J. The antiphospholipid syndrome. The New England Journal of Medicine. 2002; **346**(10):752-763. DOI: 10.1056/ NEJMra002974

[73] Giannakopoulos B, Passam F, Ioannou Y, Krilis SA. How we diagnose the antiphospholipid syndrome. Blood. 2009;**113**(5):985-994. DOI: 10.1182/blood-2007-12-129627. Epub 2008 Aug 28

[74] Kim Y, Kim SY. Antiphospholipid antibody and recurrent ischemic stroke: A systematic review and meta-analysis. Stroke. 2020;**51**(12):3728-3732

[75] Cohen D, Berger SP, Steup-Beekman GM, Bloemenkamp KW, Bajema IM. Diagnosis and management of the antiphospholipid syndrome. BMJ. 2010;**340**:c2541

[76] Giron-Gonzales JA, Garcia del Rio E, Rodriguez C, Rodriguez-Martorell J, Serrano A. Antiphospholipid syndrome and asymptomatic carriers of antiphospholipid antibody: Prospective analysis of 404 individuals. The Journal of Rheumatology. 2004;**31**:1560-1567

[77] Pengo V, Testa S, Martinelli I, Ghirarduzzi A, Legnani C, Gresele P, et al. Incidence of a first thromboembolic event in carriers of isolated lupus anticoagulant. Thrombosis Research. 2015;**135**(1):46-49

[78] Pengo V, Ruffatti A, Legnani C, Testa S, Fierro T, Marongiu F, et al. Incidence of a first thromboembolic event in asymptomatic carriers of high-risk antiphospholipid antibody profile: A multicenter prospective study. Blood. 2011;**118**(17):4714-4718

[79] Arnaud L, Mathian A, Ruffatti A, Erkan D, Tektonidou M, Cervera R, et al. Efficacy of aspirin for the primary prevention of thrombosis in patients with antiphospholipid antibodies: An international and collaborative meta-analysis. Autoimmunity Reviews. 2014;**13**(3):281-291

[80] Cuadrado MJ, Bertolaccini ML, Seed PT, Tektonidou MG, Aguirre A, Mico L, et al. Low-dose aspirin vs low-dose aspirin plus low-intensity warfarin in thromboprophylaxis: A prospective, multicentre, randomized, open, controlled trial in patients positive for antiphospholipid antibodies (ALIWAPAS). Rheumatology. 2014;**53**(2):275-284

[81] Tektonidou MG, Andreoli L, Limper M, Amoura Z, Cervera R, Costedoat-Chalumeau N, et al. EULAR recommendations for the management of antiphospholipid syndrome in adults. Annals of the Rheumatic Diseases. 2019;**78**(10):1296-1304

[82] Baigent C, Blackwell L, Collins R, Emberson J, Godwin J, Peto R, et al. Aspirin in the primary and secondary prevention of vascular disease: Collaborative meta-analysis of individual participant data from randomised trials. Lancet. 2009;**373**(9678):1849-1860

[83] Amory CF, Levine SR, Brey RL, Gebregziabher M, Tuhrim S, Tilley BC, et al. Antiphospholipid antibodies and recurrent thrombotic events: Persistence and portfolio. Cerebrovascular Diseases. 2015;**40**(5–6):293-300

[84] Crowther MA, Ginsberg JS, Julian J, Denburg J, Hirsh J, Douketis J, et al. A comparison of two intensities of warfarin for the prevention of recurrent thrombosis in patients with the antiphospholipid antibody syndrome. New England Journal of Medicine. 2003;**349**(12):1133-1138

[85] Finazzi G, Marchioli R, Brancaccio V, Schinco P, Wisloff F, Musial J, et al. A randomized clinical trial of high-intensity warfarin vs. conventional antithrombotic therapy for the prevention of recurrent thrombosis in patients with the antiphospholipid syndrome (WAPS) 1. Journal of Thrombosis and Haemostasis. 2005;**3**(5):848-853

[86] Cohen H, Isenberg DA. How I treat anticoagulant-refractory thrombotic antiphospholipid syndrome. Blood. 2021;**137**(3):299-309

[87] Lim W, Crowther MA, Eikelboom JW. Management of antiphospholipid antibody syndrome: A systematic review. Journal of the American Medical Association. 2006;**295**(9):1050-1057

[88] Andrade D, Cervera R, Cohen H, Crowther M, Cuadrado MJ, Canaud G, et al. 15th international congress on antiphospholipid antibodies task force on antiphospholipid syndrome treatment trends report. In: Antiphospholipid Syndrome. Champions: Springer; 2017. pp. 317-338

[89] Cohen H, Hunt BJ, Efthymiou M, Arachchillage DR, Mackie IJ, Clawson S, et al. Rivaroxaban versus warfarin to treat patients with thrombotic antiphospholipid syndrome, with or without systemic lupus erythematosus (RAPS): A randomised, controlled, open-label, phase 2/3, non-inferiority trial. The Lancet Haematology. 2016; **3**(9):e426-e436

[90] Pengo V, Denas G, Zoppellaro G, Jose SP, Hoxha A, Ruffatti A, et al. Rivaroxaban vs warfarin in high-risk patients with antiphospholipid syndrome. Blood. 2018;**132**(13): 1365-1371

[91] Erkan D, Willis R, Murthy VL, Basra G, Vega J, Ruiz-Limón P, et al. A prospective open-label pilot study of fluvastatin on proinflammatory and prothrombotic biomarkers in antiphospholipid antibody positive patients. Annals of the Rheumatic Diseases. 2014;**73**(6):1176-1180

[92] Wallace DJ, Linker-Israeli M, Metzger AL, Stecher VJ. The relevance of antimalarial therapy with regard to thrombosis, hypercholesterolemia and cytokines in SLE. Lupus. 1993;**2** (Suppl 1):S13-S15

[93] Kravvariti E, Koutsogianni A, Samoli E, Sfikakis PP, Tektonidou MG. The effect of hydroxychloroquine on thrombosis prevention and antiphospholipid antibody levels in primary antiphospholipid syndrome: A pilot open label randomized prospective study. Autoimmunity Reviews. 2020; **19**(4):102491

[94] Chighizola CB, Ubiali T, Meroni PL. Treatment of thrombotic antiphospholipid syndrome: The rationale of current management—An insight into future approaches. Journal of Immunology. Research. 2015;**2015**: 951424

[95] Bazán EC, López CD, Lozano PM, Cervera R, Espinosa G. Discontinuation of anticoagulation or antiaggregation treatment may be safe in patients with primary antiphospholipid syndrome when antiphospholipid antibodies became persistently negative. Immunologic Research. 2013;**56**(2–3): 358-361

[96] Criado-García J, Fernández-Puebla RA, Jiménez LL, Velasco F, Santamaría M, Blanco-Molina A. Anticoagulation treatment withdrawal in primary antiphospholipid syndrome when anticardiolipin antibodies become negative. Revista Clinica Espanola. 2008;**208**(3):135-137

[97] Bertsias GK, Ioannidis JP, Aringer M, Bollen E, Bombardieri S, Bruce IN, et al. EULAR recommendations for the management of systemic lupus erythematosus with neuropsychiatric manifestations: Report of a task force of the EULAR standing committee for clinical affairs. Annals of the Rheumatic Diseases. 2010;**69**(12):2074-2082

[98] Ageno W, Crotti S, Turpie AG. The safety of antithrombotic therapy during pregnancy. Expert Opinion on Drug Safety. 2004;**3**(2):113-118. DOI: 10.1517/eods.3.2.113.27343

Chapter 3

Obstetric Antiphospholipid Syndrome

Ariela Hoxha and Paolo Simioni

Abstract

Antiphospholipid syndrome (APS) is characterized by thrombotic events and obstetric complications in the presence of persistently positive antiphospholipid antibodies. Obstetric manifestations include, recurrent miscarriages, fetal death at or beyond the 10th week of gestation, and premature birth due to pre-eclampsia/placental insufficiency. Even now, both clinical features and laboratory parameters are controversial. Both can be used to stratify women with APS in terms of risk of adverse pregnancy outcome, and thus adjust treatment. APS pregnancies should be classified into low, medium and high-risk classes based on clinical and laboratory features. Depending on the risk class, the most appropriate therapy must be then selected. Heparin plus LDA is considered the standard of care for patients with a confirmed diagnosis of obstetric APS and generally results in over 70–80% successful pregnancies. The 20–30% pregnancies in which treatment fails are defined as "high-risk" or "refractory" pregnancies. Numerous treatments have been used in addition to standard of care, to treat these patients, but no well-designed trial has yet been conducted. New insights into the etiopathogenetic mechanisms of obstetric APS have led to the testing of new therapeutic approaches, that may soon change the way we manage this condition.

Keywords: fetal death, obstetric antiphospholipid syndrome, antiphospholipid antibodies, pregnancy, therapy

1. Introduction

Antiphospholipid syndrome (APS) is a rare systemic autoimmune disease characterized by thrombotic events and obstetric complications in the presence of persistently positive antiphospholipid antibodies (aPLs) [1]. The condition may occur alone, that is primary APS, or in association with other autoimmune diseases, most commonly systemic lupus erythematosus (SLE), and is then referred to as secondary APS. The classification criteria (**Table 1**), developed in 1999 [2] and revised in 2006 [1] include clinical features consisting of thrombosis and/or obstetric morbidity in the presence of laboratory criteria such as lupus anticoagulant (LA), medium-high titer IgG/IgM anticardiolipin antibodies (aCL) and/or anti-β2 glycoprotein I (anti-β2GPI). They are often used, also, as diagnostic tools. Obstetric APS (OAPS) subsets are featured by recurrent early miscarriages, fetal death at or beyond 10 weeks of gestation, and early delivery due to severe preeclampsia or placental insufficiency [3, 4]. The first associations between recurrent pregnancy loss and a circulating anticoagulant later known as LA, date back to 1975 [5], but it was not until 1984 Hughes linked the presence of aCL with recurrent miscarriages

IntechOpen

defining APS [6]. Nowadays, OAPS is considered one of the few treatable causes of recurrent loss and represents an important health burden for women of childbearing age and a challenge for the physicians [7]. Management of OAPS is challenging for the physician, as individual women with APS do not have the same obstetric risk profile. In the last decade, the importance to stratifying them based on their laboratory and clinical features has been emphasized to quantify the risk of adverse pregnancy outcome. Many efforts have also been made to adjust therapy according to risk stratification [8]. Moreover, new insights into the pathogenesis and clinical understanding of APS have led to potential new therapeutic approaches [9, 10].

This chapter aims to clarify aspects of pathogenesis, clinical features, risk stratification and therapeutic strategies in OAPS.

2. Antiphospholipid antibodies

The aPLs are a heterogeneous group of autoantibodies that bind primarily to circulating plasma proteins such as β2GPI, prothrombin, and others when bound to phospholipids themselves. The prevalence of aPL is about 1–5% in the general population and increase up to 40% in patients with pregnancy complications [11–14].

2.1 Criteria antiphospholipid antibodies

Currently, there are three types of aPLs, depending on the detection method, which are included in the laboratory classification criteria for APS (**Table 1**).

Clinical criteria
• *Vascular thrombosis*
One or more clinical episodes of arterial, venous or small vessels thrombosis in any organ or tissue
Thrombosis must be confirmed by appropriate imaging studies or histopathology
Thrombosis on histopathology specimen must be present without inflammation of vessel wall
• *Obstetric morbidity*
One or more unexplained fetal death of a morphologically normal fetus at or beyond 10th week of gestation
One or more premature birth of a morphologically normal neonate before 34th week of gestation due to placental insufficiency and/or eclampsia or severe pre-eclampsia
Three or more consecutive early miscarriages before 10th week of gestations; with maternal and paternal factors such as anatomical, hormonal and chromosomal abnormalities should be ruled out
Laboratory criteria
Medium to high levels of aPLs antibodies detected on two or more occasions, at least 12 weeks apart
• Lupus anticoagulant, detected according to the guidelines of the International Society on Thrombosis and Hemostasis
• Anti-cardiolipin antibodies (IgG and/or IgM) at medium-high levels (>40 units or above the 99th percentile)
• Anti-B2-glycoprotein I antibodies IgG and/or IgM at medium-high levels (>40 units or above the 99th percentile)
aPLs: antiphospholipid antibodies, IgG: immunoglobulin G, IgM: immunoglobulin M.

Table 1.
Clinical and laboratory criteria of definite antiphospholipid syndrome. APS is diagnosed when at least one of the following clinical criteria and one of the following laboratory criteria are met.

2.1.1 Lupus anticoagulant

LA are heterogenous antibodies detected with a functional test that measures the ability of aPL to prolong phospholipid-dependent clotting reactions. Anti-β2GPI [15] and anti-prothombin (aPT) antibodies [16] have been identified as the main mediators of this reaction. LA detection is very challenging, as it has many pitfalls leading to either false positive or false negative results. The International Society for Thrombosis and Hemostasis (ISTH) guidelines released in 2009, updated in 2018 provided a step toward standardization of LA [17, 18]. Following those guidelines LA detection is based on the simultaneously use of two assays with different principles following a multi-step procedure, with screening, mixing and confirmation steps. The most commonly used are activated partial thromboplastin time followed by the diluted Russell's viper venom test. The presence of LA should always be confirmed by performing the assays in the presence of excess of phospholipids, with a correction of the prolongation of the times as a result.

2.1.2 Anticardiolipin antibodies

aCL are heterogeneous antibodies that in immunoassay's bind to a complex of phospholipids and plasma proteins, mainly β2GPI. In this assay there can be measured two types of aCL, "β2GPI-indipendent" that bind to phospholipids alone and are typically free of clinical significance and "β2GPI-dipendent" which are related to clinical manifestations of APS [19–21].

2.1.3 Anti-β2 glycoprotein antibodies

Anti-β2GPI antibodies are specific to the β2GPI, a cofactor with affinity for anionic phospholipids which inhibits in vitro the activation of prothrombin and the ADP-dependent platelet aggregation [22].

Both aCL and anti-β2GPI antibodies of IgM and IgG isotypes are detected by immunological assays following the ISTH subcommittee recommendation [23]. IgG and IgM isotype, at medium-high titer have greater clinical significance [24].

2.2 Non-criteria antiphospholipid antibodies

A series of autoantibodies non included in laboratory classification criteria the so called "non-criteria" aPLs have been reported in the last years related to APS manifestations. Those directed against two major phospholipids-binding protein representing the true antigenic targets for aPL (i.e. prothrombin and β2GPI) have demonstrated the highest significant association with thrombotic and obstetric features of APS [25–27].

2.2.1 Anti-prothrombin antibodies

APT antibodies are detected by ELISA using a purified prothrombin as antigen coated onto irradiated plates [28] or phosphatidylserine/prothrombin complex [29]. Although a correlation between the two assays have been reported, these antibodies differed either in affinity or in epitopes that they recognized [30]. The ones directed against anti-phosphatidylserine/prothrombin complex (aPS/PT) seems having a closer association with APS and LA activity than with antibodies to prothrombin alone [31, 32]. aPS/PT have been reported to be significantly associated with both thrombotic and obstetric manifestations of APS [33–35]. Moreover, since they have been shown to be closely related to LA, have been proposed as a surrogate test for and as an additional serologic marker of APS, to be performed with other aPL tests to improve diagnosis [36, 37].

2.2.2 Anti-domain I antibodies

A subgroup of anti-β2GPI, those directed to domain I of the molecule [15], have been reported to be strongly associated with thrombosis and LA in APS patients while those directed to domain IV/V are less frequent [25, 26]. Recently has been suggested that the ratio between anti-β2GPI-DI and anti-β2GPI-DIV/V IgG can provide a better profile of anti-β2GPI antibodies linked to APS and antibodies occurring in other pathologic condition [38].

To improve risk prediction of recurrent thrombosis and pregnancy loss the Global Anti-Phospholipid Syndrome Score (GAPSS) was developed, considering the aPL profile, conventional cardiovascular risk factors, and autoimmune antibody profile. Validated in APS and in SLE patients, a high GAPSS score predicted thrombosis better than aPLs alone [39, 40]. Recently, the GAPSS score has been shown to be a useful tool for predicting a higher likelihood of favorable pregnancy outcome in pregnant women treated with conventional therapy [41].

3. Pathogenesis of obstetric antiphospholipid syndrome

The exact etiopathogenetic mechanism liable for obstetric morbidity in APS is not yet known. The aPLs are not only a diagnostic marker but have a key role in determining thrombosis and obstetric complications [42]. In the early stages, during pregnancy, the cytotrophoblastic cells differentiates into two cell types. The villous trophoblast will fuse to form the syncytiotrophoblast, a barrier of protection between the mother and the fetus. While, the extravillous trophoblast will progressively invade and colonize the maternal endometrium [43]. aPLs target the placenta, especially the cytotrophoblastic cells. Trophoblast, synthetize β2GPI, a 70 kDa cationic protein that is normally in a "closed conformation", when free in the plasma of patients. It is composed of five homologous domains of approximately 60 amino acids each. Domains I and V are the two domains positively charged [44]. During normal pregnancy and syncytiotrophoblast formation, anionic phospholipids are externalized at trophoblastic cell surface, leading to the binding of β2GP1 via domain V. This binding offers a potential site of actions for aPL by changing the conformation of the protein from a circular to an open form and exposing domains I–IV to the surface [45, 46]. aPL have been incriminated in alteration of trophoblastic cells via different mechanisms. Pathogenesis of aPL in pregnancy include thrombotic mechanisms, inflammation, apoptosis and immunomodulatory molecules impairments in trophoblast [47].

3.1 Thrombotic mechanisms

The placental infarctions due to aPLs-mediated thrombosis of spiral arteries have been thought to be the main cause of fetal demise [48]. However, thrombosis of placental surface is not a universal feature. Recently, placental infarction has been demonstrated [49], only in one third of the placenta of aPLs-positive women and moreover, similar lesions were also reported in those of aPLs-negative women who had had a miscarriage [50, 51]. According to a review of 34 studies comparing the prevalence of placental features between aPL-positive women and aPLs-negative ones, five lesions, have been identified, as fingerprint of human placental aPLs including: placental infarction, impaired spiral artery remodeling, decidual inflammation, an increase of syncytial knots and a decrease of vasculo-syncytial membranes [52]. These different features of aPLs placenta fingerprints give rise to thought that pregnancy complications by aPLs are due to different pathologic events mainly non-thrombotic related.

3.2 Non-thrombotic mechanisms

The non-thrombotic mechanism which leads to defective placentation are thought to be the main cause of obstetric manifestations. Especially, aPLs have direct effect on trophoblast viability and syncytialization as well as on trophoblast invasion and alter the production of syncyotrophoblast hormones. Moreover, signs of inflammation within the decidua, such as fibrin deposits, were more represented than thrombosis in histological analysis of placenta from women with APS, suggesting another mechanism in pregnancies affected by aPL [52, 53].

3.2.1 Defective placentation

aPLs affect trophoblast viability by both decreasing their proliferation and promoting their death [49] and by altering their expression of the apoptotic regulators BcL-2 and Bax [54]. Furthermore, aPLs decrease the expression of caspases 3 and 7, suggesting that they are involved in death mechanism of the trophoblast [55]. Moreover, as demonstrated by different studies [49, 56], aPLs are involved in the inhibition of syncytialization, an essential process for the replenishment of the syncytiotrophoblast. It has been speculated that the mechanism by which aPLs inhibit syncytializationis due to the decrease of caspase expression, the activation of which is required for cytotrophoblast fusion [52]. Thus, proliferation of cytotrophoblast is reduced, while death increases. This results in a fewer cytotrofoblasts available to replenish the syncytiotrophoblasts. On the other hand, increased death of the syncytiotrofoblasts, leads to increased production of trophoblast debris and increased denudation of syncytiotrophoblasts and fibrinoid deposition. This process leads to a decrease in de syncytialisation and thus impaired placentation [49].

3.2.2 Trophoblast invasion

Trophoblast invasion, into maternal spiral arteries, is an essential physiologic change that allows the anchoring of placenta to the decidua as well as the transplacental passage of nutrients and wastes between the mother and the fetus. Several studies in vitro [49] have shown that aPLs reduce the ability of extra-villous cytotrophoblast to invade the maternal decidua, so affecting both the anchorage of placenta and the spiral arteries transformation, the latter leading to a reduced blood flow to the placenta. The aPLs are thought to impaire trophoblast invasion by altering the expression of adhesion molecules such as placental growth factor (PIGF), vascular endothelial growth factor (VEGF) and soluble FmS-like kinase I (sFlt-I) as well as cytokines such as interleukin 1β [49, 52, 57]. Altered trophoblast invasion of leads to impaired transplacental passage resulting in pre-eclampsia and intrauterine growth restriction (IUGR). In fact, increased sFlt-I levels in the first trimester have been shown to correlate with later onset pre-eclampsia, suggesting them as predictors of preeclampsia [58].

3.2.3 Inflammation, complement activation, and disruption of annexin shield

A well-known fingerprint of APS placenta histology is inflammation. In addition to increased release of cytokines such as IL-1β as described above, aPLs effect complement activation. Girardi et al. [59] had shown that aPLs increased complement deposition on the trophoblast surface in vitro. While, murine models of APS demonstrate a crucial role of the complement system in determining pregnancy morbidity [59–61], on the other hand, the placenta of women with APS showed deposition of C4d and C3b [62]. Moreover, data from the PROMISSE study in pregnant SLE and/or APS or aPL positive patients, showed that detection of increased Bb and sC5b-9 levels early in pregnancy was predictive of adverse pregnancy outcome, confirming complement activation as a contributor to pregnancy failure [63]. Complement activation stimulates neutrophils to release tumor necrosis factor-alfa (TNF-α); pregnant mice lacking TNF-α are protected from pregnancy loss induced by injections of aPLs [64].

Last, but not least, aPLs disrupt the binding of annexin V, an anticoagulant protein that crystallizes over phospholipid bilayers blocking their availability for coagulation reactions. This disruption additionally contributes to both thrombosis and miscarriages in the APS [65, 66].

Overall, the pathogenetic mechanisms by which aPLs cause obstetric complications are complex and include both inflammatory and non-inflammatory mechanisms, which are not mutually exclusive and may coincide in time. This could reflect the different characteristics of fetal complications. Adequate invasion of the trophoblast into the maternal decidua remains crucial for a normal-evolving pregnancy. Inadequate invasion of the maternal spiral arteries by the extravillous cytotrophoblast and severe inflammation of the placenta, together with reduced hormone secretion leading to early miscarriages are thought to be the major pathogenic mechanisms of early pregnancy loss in APS patients. While, a lack of transformation of the maternal spiral arteries together with activation of the complement and of the coagulation cascade are responsible for late pregnancy loss and preeclampsia.

4. Clinical manifestations of obstetric APS

Obstetric morbidity of APS is characterized by various pregnancy complications such as recurrent miscarriage, fetal death, and premature birth. These manifestations can occur in the same patient during her childbearing years.

Recurrent early miscarriage (REM) is the most frequent obstetric feature of APS. In the European Registry on Obstetric Antiphospholipid Syndrome REM was observed in almost 54% of women with obstetric APS [67]. On the other hand, aPLs are found in up to 20% of women who experience an early abortion [11]. REM can be caused by various maternal and paternal factors such as anatomical abnormalities, endocrine diseases such as diabetes and thyroiditis, autoimmune diseases, parental chromosomal abnormalities and infectious agents. Therefore, these causes must be systematically excluded specifically in cases where REM is the only clinical manifestation.

In contrast, fetal death and premature birth due to pre-eclampsia and/or placental insufficiency are considered more specific clinical manifestations of APS [3]. Fetal death, in particular the late fetal loss, i.e. beyond 20 weeks of gestation is strongly associated with aPLs [68, 69]. In a population-based study by The Stillbirth Collaborative Research Network in the United States [69], elevated aCL and anti-β2GPI levels were associated with a 3- to 5-fold increased likelihood of stillbirth. Pre-eclampsia, eclampsia, and placental abruption are maternal complications of APS. Preeclampsia occurs in approximately 10–17% of pregnancies with APS, compared to 3–5% of pregnancies without the condition [3, 70]. Preeclampsia in patients with APS is often severe and occurs early in pregnancy [70]. These data were recently confirmed in a case-control study, which found that more than 10% of women who gave birth before 34 weeks' gestation due to pre-eclampsia or placental insufficiency were positive for aPLs [71]. In a study of 1000 consecutive APS pregnancies, the presence of early pre-eclampsia and early IUGR was found in 181 (18.1%) and 161 (16.1%), respectively, despite treatment [67].

Even when treated with heparin plus low-dose aspirin (LDA), 9–10% of pregnant women with APS develop pre-eclampsia. This highlights the fact that counseling these pregnancies and identifying risk factors are very important to personalize therapy, as will be discussed later.

5. Risk stratification in obstetric antiphospholipid syndrome

The outcome of pregnancy in women with APS depends on their clinical history and aPLs profile. Therefore, women with APS or high levels of aPLs should be counseled before pregnancy to perform risk stratification.

Several risk factors are predictors of poor pregnancy outcome. The presence of triple aPL positivity [72, 73], which refers to IgG/IgM aCL plus IgG/IgM anti-β2GPI plus LA, correlates strongly with vascular thrombosis and pregnancy morbidity. Moreover, the presence of persistent positive LA [74, 75] has been reported as the strongest predictor for either pregnancy loss or recurrent thrombosis. These high-risk aPL profiles (**Table 2**) are associated with an increased risk of pregnancy morbidity such as intrauterine growth restriction and premature birth as well as pre-eclampsia despite appropriate anticoagulant treatment [73, 76–78]. Regarding clinical features, women with aPLs and a history of thrombosis, severe pregnancy complications such as pre-eclampsia, eclampsia, or HELLP syndrome, a concomitant SLE diagnosis, or low complement levels are associated with a higher risk of pregnancy morbidity [78, 79]. On the other hand, there are consistent data showing that a history of pregnancy morbidity alone and a single aPLs positivity for aCL or anti-β2GPI are associated with a higher live birth rate [73].

According to the risk stratification, women with obstetric APS can be divided into three groups (**Figure 1**), namely A-low risk pregnancy, those with obstetric morbidity alone and single or double positivity for aPL, B-medium risk pregnancy, those with prior thrombosis and single or double positivity for aPL, and C-high risk pregnancy, those with prior thrombosis and/or severe pregnancy complications and LA/triple positivity for aPL. This subdivision could guide the therapeutic approach in these patients.

Pregnant women with APS must also be closely monitored during pregnancy to promptly identify signs of placental insufficiency. It has been shown that upgrading therapy at the first signs of placental insufficiency results in a higher birth rate [80].

High-risk aPLs profiles
• Persistent LA positivity (measured according to ISTH guidelines in two or more occasions at least 12 weeks apart)
• Double aPL positivity (any combination of lupus anticoagulant, aCL antibodies or anti-β2 glycoprotein I antibodies)
• Triple aPL positivity (LA + aCL IgG/IgM + anti-β2GPI IgG/IgM
Low-risk aPLs profiles
• Isolated, positive aCL or anti-β2GPI IgG/IgM at low-medium titres, particularly if transiently positive

aPLs: antiphospholipid antibodies, LA: lupus anticoagulant, aCL: anti-cardiolipin antibodies, anti-β2GPI: anti-β2glycoprotein I, IgG: immunoglobulin G, IgM: immunoglobulin M.

Table 2.
Definition of high-risk and low-risk antiphospholipid antibodies profile.

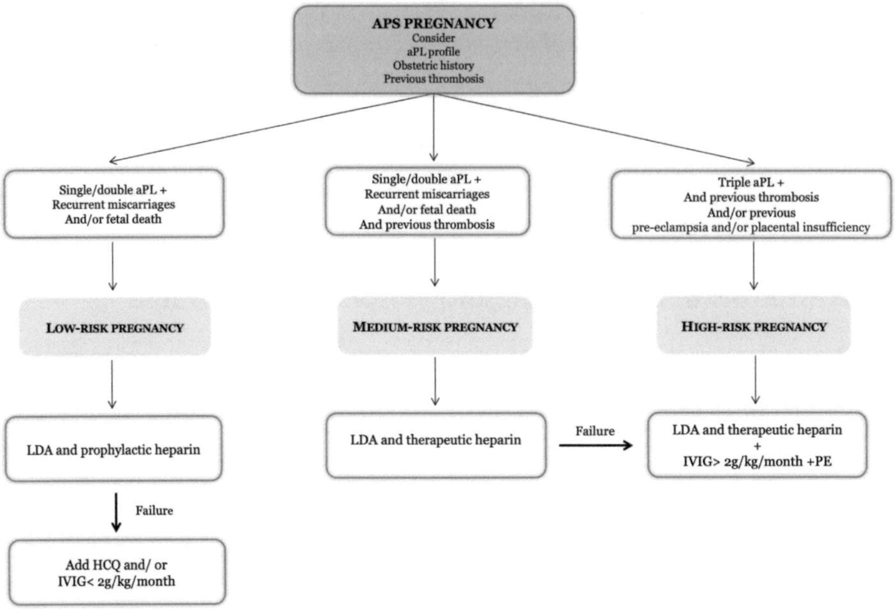

Figure 1.
Risk stratification and management of antiphospholipid pregnancies. aPL: antiphospholipid antibodies, LDA: low dose aspirin; IVIG: intravenous immunoglobulin; PE: plasma-exchange; HCQ: hydroxychloroquine.

Abnormal uterine artery flow on Doppler ultrasound is an indirect indicator of the development of placental insufficiency and/or pre-eclampsia [81]. Therefore, the European Alliance of Associations for Rheumatology, (EULAR) guidelines recommend the use of uterine artery Doppler ultrasonography for the management of SLE/APS patients [82]. In addition, a drop in platelet count in the first trimester has recently been found to be associated with the development of pre-eclampsia in APS pregnancy compared to non-APS pregnancies. APS women who later developed pre-eclampsia and/or placental insufficiency had a decrease in initial platelet count of more the 20% (personal observation, not published).

6. Management of the obstetric antiphospholipid syndrome

The current standard of care (**Table 3** and **Figure 1**) for the management of APS pregnancies [83], although controversial and supported by only a limited number of well-designed studies, is the prophylactic administration of heparin plus LDA for individuals with pregnancy morbidity alone. Mothers with a history of thrombosis alone or in association with pregnancy morbidity are usually treated with therapeutic heparin in combination with LDA to prevent both thrombosis and pregnancy morbidity. The data supporting these recommendations come almost exclusively from clinical trials evaluating the prevention of recurrent early miscarriages rather than late pregnancy complications. A total of 140 women with APS-related recurrent early miscarriages, were enrolled in two randomized control trials comparing treatment with LDA alone or in combination with heparin [84, 85]. The combination of LDA with heparin showed a significantly higher live birth rate than LDA alone. These data were however not confirmed in two subsequent studies [86, 87]. Nevertheless, a subsequent follow-up meta-analysis [88] and a recent Cochrane review [89] concluded that combining heparin with LDA during pregnancy may increase the live birth rate in women with APS compared with LDA alone. Since low

Clinical/laboratory features	Treatment
High-risk aPLs profile but no history of thrombosis or pregnancy complications, with or without SLE.	Close monitoring of mother and fetus; LDA (75–100 mg daily) during pregnancy should be considered.
Obstetric APS only (no prior thrombotic events), with or without SLE:	
a. History of ≥3 recurrent spontaneous miscarriages <10th week of gestation or history of fetal loss (≥10th week of gestation),	LDA and heparin at prophylactic dosage.
b. History of delivery <34 weeks of gestation due to eclampsia or severe pre-eclampsia or due to recognized features of placental insufficiency,	LDA or LDA and heparin at prophylactic dosage considering the individual's risk profile.
c. Clinical 'non-criteria' obstetric APS such as the presence of two recurrent spontaneous miscarriages <10th week of gestation, or delivery ≥34 weeks of gestation due to severe pre-eclampsia or eclampsia.	LDA alone or in combination with heparin might be considered based on the individual's risk profile.
Thrombotic APS.	LDA and therapeutic dose of heparin.
Patients with obstetric APS with recurrent pregnancy complications despite combination treatment with LDA and heparin at prophylactic dosage.	Increasing heparin dose to therapeutic dose; consider addition of HCQ or low-dose prednisolone in the first trimester and IVIG in highly selected cases.

APS: antiphospholipid syndrome, aPLs: antiphospholipid antibodies, LDA: low dose aspirin, HCQ: hydroxychloroquine, IVIG: intravenous immunoglobulin, SLE: systemic lupus erythematosus.

Table 3.
Current recommendation [83] for the management of pregnant women with antiphospholipid antibodies or APS.

molecular weight heparin (LMWH) is easier to administer and have less adverse events it is the drug of choice in most cases. Furthermore, the dose of LMWH should be personalized. Case-control studies comparing a fixed dose of LMWH with a weight-adjusted dose of LMWH have shown a higher live birth rate with the latter [80, 90]. Several studies, recently summarized [91] suggest that women with either clinical and/or laboratory non-criteria manifestations of obstetric APS may benefit from standard obstetric APS treatment with LMWH plus LDA, with good pregnancy outcomes.

6.1 Management of refractory/high-risk pregnancies

Even if current recommendations are carefully followed [83], 20–30% of pregnancies fail [92], and these are the so-called refractory pregnancies and/or high-risk pregnancies. High-risk pregnancies are pregnancies of APS patients with one or more laboratory or clinical risk factors who may or may not have experienced adverse pregnancy outcomes despite treatment with heparin/LDA [8]. Experts in the field believe that women should receive additional treatments when the risk of having pregnancy complications is elevated based on their antibody profile and certain clinical characteristics [93] as this will improve these women's live birth rate and/or reduce their pregnancy complications. Various therapeutic options, such as low-dose prednisolone, intravenous immunoglobulins, hydroxychloroquine, plasmapheresis alone or in combination have been used in an attempt to achieve a better pregnancy outcome [94–104]. Usually, these treatments were administered in conjunction with conventional heparin/LDA therapy. An Experts' Consensus [8] following a systematic review of the literature recently suggested that hydroxychloroquine and low-dose steroids, alone or in combination, may be an option for pregnant APS women whose previous pregnancies were not successful despite receiving conventional therapy. Intravenous immunoglobulins (IVIG) and plasma exchange (PE), alone or in combination, could be considered in refractory high-risk APS pregnancies. Furthermore, a recent systematic literature review [9] analyzed 313 refractory/high-risk pregnancies from 14 studies comprising 134 (42.8%) pregnancies refractory to conventional treatment, and 179 (57.2%) high-risk/refractory pregnancies. The findings from this review suggest introducing low-dose IVIG (< 2 g/kg/month) or HCQ 400 mg/day before pregnancy in women with APS refractory to conventional therapy, and high-dose IVIG (2 g/kg/month) in combination with PE or alone in those with high risk/refractory APS with both approaches leading to improved pregnancy outcome. It should be noted that, drug related side-effects were observed in 3/313 (0.9%) of pregnancies, and none of which required hospitalization.

Although statins appear to be a potential therapy in the treatment of APS refractory pregnancies, as suggested by murine studies [105] and a small case-control study of the use of pravastatin for placental dysfunction/pre-eclampsia in patients with APS [106], they have not yet been routinely used to date. Pravastatin was administered to 11 women with obstetric APS and preeclampsia/IUGR at the time of diagnosis of the complication (range 22–30 weeks) in addition to standard of care; they were compared with 10 control patients with preeclampsia/IUGR who did not receive pravastatin. The pravastatin group achieved a 100% rate of live births (34–36 week of gestation) and rapid improvement in uterine artery hemodynamics was observed, while the control group had a 50% rate of live births, all of which were delivered preterm (26–32 week of gestation). An ongoing multicenter study, the StAmP trial, is testing pravastatin for the prophylaxis of preeclampsia (double-blind, randomized and placebo-controlled) in the general population [107]. However, several concerns remain about their use, as they are classified as FDA category X. However, no congenital abnormalities have been reported in the pilot studies to date.

7. Future perspectives

Much has been done over the past three decades to understand the pathogenesis of aPL-mediated obstetric injury and to diagnose and treat obstetric APS. However, much remains to be done to provide the best diagnostic and therapeutic approach to our patients.

Future research should be conducted to evaluate the intracellular signaling pathways that are affected by aPLs and lead to trophoblastic dysfunction. Identification of these mechanisms could lead to identification of potential therapeutic targets in the future.

The redefinition of the classification criteria is currently under evaluation and should provide a valuable tool for future clinical trials of APS. It is important to include in these studies the concept of risk stratification according to aPLs profile and clinical features. The possible inclusion of new autoantibodies such as anti-domain I and aPS/PT in the new classification criteria could be helpful in improving the diagnosis of obstetric APS, especially in cases where conventional antibodies are not detectable.

Clinical trials of new therapeutic approaches for refractory obstetric APS syndrome are currently underway. Two randomized clinical trials (NCT04275778 and NCT04624269) and two prospective studies [108, 109] are evaluating the effect of HCQ in addition to standard treatment to prevent pregnancy morbidity in APS patients. According to a recent prospective case series, the use of TNF-alpha inhibitors in addition to standard treatment seems to be a promising treatment for refractory obstetric APS [10]. If these findings are confirmed by the ongoing IMPACT Study: IMProve Pregnancy in APS with certolizumab (NCT03152058), the TNF-alfa inhibitors may constitute a valid second-line treatment for refractory and/or high-risk APS pregnancies in the near future.

8. Conclusions

Nowadays, we have gained new insights into the pathogenesis and management of obstetric APS. Contrary to what was first thought, aPLs determine pregnancy morbidity with both inflammatory and non-inflammatory mechanisms. These findings have led to a better understanding of the different features of obstetric APS. While, inadequate invasion of maternal spiral arteries by the extravillous cytotrophoblast leads to early miscarriage, a lack of transformation of these arteries along with activation of the complement and the coagulation cascade is responsible for late pregnancy loss and preeclampsia. APS pregnancies should be classified into low, medium and high-risk classes based on clinical and laboratory features. Depending on the risk class, the most appropriate therapy must then be selected. Although studies have shown that intervention at the first signs of placental insufficiency can improve the pregnancy outcome, it is advisable to initiate the most appropriate therapy based on the risk class immediately at the beginning of pregnancy. It should be remembered that invasion of the trophoblast into the maternal spiral arteries occurs in the very early stages of placentation and adequate anchorage of the placenta is essential for the development of the pregnancy. Therefore, we need to start most appropriate therapy as soon as possible to facilitate a favorable pregnancy outcome.

Conflict of interest

The authors declare no conflict of interest.

Author details

Ariela Hoxha[1*] and Paolo Simioni[2]

1 Internal Medicine Unit, Department of Medicine, San Bortolo Hospital, Vicenza, Italy

2 General Internal Medicine and Thrombotic and Hemorrhagic Unit, University of Padua, Italy

*Address all correspondence to: arielahoxha@hotmail.com

IntechOpen

© 2022 The Author(s). Licensee IntechOpen. This chapter is distributed under the terms of the Creative Commons Attribution License (http://creativecommons.org/licenses/by/3.0), which permits unrestricted use, distribution, and reproduction in any medium, provided the original work is properly cited.

References

[1] Miyakis S, Lockshin MD, Atsumi T, Branch DW, Brey RL, Cervera R, et al. International consensus statement on an update of the classification criteria for definite antiphospholipid syndrome (APS). Journal of Thrombosis and Haemostasis. 2006;**4**:295-306. DOI: 10.1111/j.1538-%207836.2006.01753

[2] Wilson WA, Gharavi AE, Koike T, Lockshin MD, Branch WD, Piette JC, et al. International consensus statement on preliminary classification criteria for definite antiphospholipid syndrome: Report of an international workshop. Arthritis & Rheumatology. 1999;**42**: 1309-1311

[3] Cervera R, Piette JC, Font J, Khamashta M, Shoenfeld Y, Camps MT, et al. Antiphospholipid syndrome: Clinical and immunologic manifestations and patterns of disease expression in a cohort of 1000 patients. Arthritis & Rheumatology. 2002;**46**:1019-1027. DOI: 10.1002/art.10187

[4] Rai RS, Regan L, Clifford K, Pickering W, Dave M, Mackie I, et al. Antiphospholipid antibodies and beta2-glycoprotein-I in 500 women with recurrent miscarriages: Results of a comprehensive screening approach. Human Reproduction. 1995;**10**:2001-2005. DOI: 10.1093/oxfordjournals.humrep.a136224

[5] Nilsson IM, Åstedt B, Hedner U, Berezin D. Intrauterine death and circulating anticoagulant ("antitromboplastin"). Acta Medicine Scandinavian. 1997;**197**:153-159. DOI: 10.1111/j.0954-6820.1975.tb04897.x

[6] Hughes GR. The Prosser-White oration 1983. Connective tissue disease and the skin. Clinical and Experimental Dermatology. 1984;**9**:535-544. DOI: 10.1111/j.1365-2230.1984.tb00856.x

[7] Schreiber K, Hunt BJ. Pregnancy and antiphospholipid syndrome. Seminars in Thrombosis and Hemostasis. 2016;**42**:780-788. DOI: 10.1055/s-0036-1592336

[8] Giacomelli R, Afeltra A, Bartoloni E, Berardicurti O, Bombardieri M, Bortoluzzi A, et al. The growing role of precision medicine for the treatment of autoimmune diseases; results of a systematic review of literature and experts' consensus. Autoimmunity Reviews. 2021;**20**:102738. DOI: 10.1016/j.autrev.2020.102738

[9] Ruffatti A, Tonello M, Favaro M, Del Ross T, Calligaro A, Ruffatti AT, et al. The efficacy and safety of second-line treatments of refractory and/or high risk pregnant antiphospholipid syndrome patients. A systematic literature review analyzing 313 pregnancies. Seminars in Arthritis and Rheumatism. 2021;**51**: 28-35. DOI: 10.1016/j.semarthrit.2020.10.001

[10] Alijotas-Reig J, Esteve-Valverde E, Llurba E, et al. Treatment of refractory poor aPL-related obstetric outcomes with TNF-alpha blockers: Maternal-fetal outcomes in a series of 18 cases. Seminars in Arthritis and Rheumatism. 2019;**49**:314-318. DOI: 10.1016/j.semarthrit.2019.02.006

[11] Lockwood CJ, Romero R, Feinberg RF, Clyne LP, Coster B, Hobbins JC. The prevalence and biologic significance of lupus anticoagulant and anticardiolipin antibodies in a general obstetric population. American Journal of Obstetrics & Gynecology. 1989;**161**: 369-373. DOI: 10.1016/0002-9378(89)90522-x

[12] Petri M. Epidemiology of antiphospholipid antibody syndrome. In: Khamashta MA, editor. Hughes Syndrome. London: Springer-Verlag; 2006. pp. 22-28

[13] Rai RS, Clifford K, Cohen H, Regan L. High prospective fetal loss rate in untreated pregnancies of women with recurrent miscarriage and antiphospholipid antibodies. Human Reproduction. 1995;**10**:3301-3304. DOI: 10.1093/oxfordjournals.humrep.a135907

[14] Andreoli L, Chighizola CB, Banzato A, Pons-Estel GJ, de Jezus GR, Erkan D. Estimated frequency of antiphospholipid antibodies in patients with pregnancy morbidity, stroke, myocardial infarction and deep vein thrombosis: A critical review of the literature. Arthritis Care and Research. 2013;**65**:1869e73. DOI: 10.1002/acr.22066

[15] de Laat B, Derksen RH, Urbanus RT, de Groot PG. IgG antibodies that recognize epitope Gly40-Arg43 in domain I of beta 2-glycoprotein I cause LAC, and their presence correlates strongly with thrombosis. Blood. 2005;**105**:1540-1545. DOI: 10.1182/blood-2004-09-3387

[16] Bevers EM, Galli M, Barbui T, Comfurius P, Zwaal RF. Lupus anticoagulant IgG's (LA) are not directed to phospholipids only, but to a complex of lipid-bound human prothrombin. Thrombosis and Haemostasis. 1991;**66**:629-632

[17] Pengo V, Tripodi A, Reber G, Rand JH, Ortel TL, Galli M, et al. Update of the guidelines for lupus anticoagulant detection. Journal of Thrombosis and Haemostasis. 2009;**7**:1737-4170. DOI: 10.1111/j.1538-7836.2009.03555.x

[18] Devreese KMJ, Ortel TL, Pengo V, de Laat B, Subcommittee on Lupus Anticoagulant/Antiphospholipid Antibodies. Laboratory criteria for antiphospholipid syndrome: Communication from the SSC of the ISTH. Journal of Thrombosis and Haemostasis. 2018;**16**:809-813. DOI: 10.1111/jth.13976

[19] Galli M, Comfurius P, Maassen C, Hemker HC, de Baets MH, van Breda-Vriesman PJ, et al. Anticardiolipin antibodies (ACA) directed not to cardiolipin but to a plasma protein cofactor. Lancet. 1990;**335**:1544-1547. DOI: 10.1016/0140-6736(90)91374-j

[20] McNeil HP, Simpson RJ, Chesterman CN, Krilis SA. Anti-phospholipid antibodies are directed against a complex antigen that includes a lipid binding inihibitor of coagulation: Beta 2-glycoprotein I (apolipoprotein H). Proceedings of the National Academy of Sciences of the United States of America. 1990;**87**:4120-4124. DOI: 10.1073/pnas.87.11.4120

[21] Hunt JE, McNeil HP, Morgan GJ, Crameri RM, Krilis SA. A phospholipid-beta 2-glycoprotein I complex is an antigen for anticardiolipin antibodies occurring in autoimmune disease but not with infection. Lupus. 1992;**1**:75-81. DOI: 10.1177/096120339200100204

[22] de Laat B, Derksen RH, van Lummel M, Pennings MT, de Groot PG. Pathogenic anti-beta2-glycoprotein I antibodies recognize domain I of beta2-glycoprotein I only after a conformational change. Blood. 2006;**107**:1916-1924. DOI: 10.1182/blood-2005-05-1943

[23] Devreese KM, Pierangeli SS, de Laat B, Tripodi A, Atsumi T, Ortel TL, et al. Testing for antiphospholipid antibodies with solid phase assays: Guidance from the SSC of the ISTH. Journal of Thrombosis and Haemostasis. 2014;**12**:792-795. DOI: 10.1111/jth.12537

[24] Kelchetermans H, Pelkmans L, de Laat B, Devreese KM. IgG/IgM antiphospholipid antibodies present in the classification criteria for the antiphospholipid syndrome: A critical review of their association with thrombosis. Journal of Thrombosis and Haemostasis. 2016;**14**:1530-1548. DOI: 10.1111/jth.13379

[25] de Laat B, Pengo V, Pabinger I, Musial J, Voskuyl AE, Bultink IE, et al. The association between circulating antibodies against domain I of beta2-glycoprotein I and thrombosis: An international multicenter study. Journal of Thrombosis and Haemostasis. 2009;7:1767-1773. DOI: 10.1111/j.1538-7836.2009.03588.x

[26] Meneghel L, Ruffatti A, Gavasso S, Tonello M, Mattia E, Spiezia L, et al. Detection of IgG anti-domain I β2-glycoprotein I antibodies by chemiluminescence immunoassay in primary antiphospholipid syndrome. Clinica Chimica Acta. 2015;**446**:201-205. DOI: 10.1016/j.cca.2015.04.033

[27] Hoxha A, Mattia E, Tonello M, Grava C, Pengo V. Ruffatti. A. Antiphosphatidylserine/prothrombin antibodies as biomarkers to identify severe primary antiphospholipid syndrome. Clinical Chemistry and Laboratory Medicine. 2017;**55**:890-898. DOI: 10.1515/cclm-2016-0638

[28] Arvieux J, Darnige L, Caron C, Reber G, Bensa JC, Colomb MG. Development of an ELISA for autoantibodies to prothrombin showing their prevalence in patients with lupus anticoagulant. Thrombosis and Haemostasis. 1995;**74**:1120-1125

[29] Atsumi T, Ieko M, Bertolaccini ML, Ichikawa K, Tsutsumi A, Matsuura E, et al. Association of autoantibodies against the phosphatidylserine-prothrombin complex with manifestations of the antiphospholipid syndrome and with the presence of lupus anticoagulant. Arthritis and Rheumatism. 2000;**43**:1982-1993. DOI: 10.1002/1529-0131(200009)43:9<1982::AID-ANR9>3.0.CO;2-2

[30] Matsuda J, Sanaka T, Nishizawa A, Gotoh M, Gohchi K. Two antiprothrombin antibodies against prothrombin and prothrombin-phosphatidyl serine show partial but not total identity. Blood Coagulation & Fibrinolysis. 2002;**8**:697-702. DOI: 10.1097/00001721-200212000-00005

[31] Hoxha A, Ruffatti A, Pittoni M, Bontadi A, Tonello M, Salvan E, et al. The clinical significance of autoantibodies directed against prothrombin in primary antiphospholipid syndrome. Clinica Chimica Acta. 2012;**413**:911-913. DOI: 10.1016/j.cca.2012.02.004

[32] Bertolaccini ML, Sciascia S, Murru V, Garcia-Fernandez C, Sanna G, Khamashta MA. Prevalence of antibodies to prothrombin in solid phase (aPT) and to phosphatidylserine-prothrombin complex (aPS/PT) in patients with and without lupus anticoagulant. Thrombosis and Haemostasis. 2013;**109**:207-213. DOI: 10.1160/TH12-07-0527

[33] Mishima S, Kaneko K, Okazaki Y, Murashima A. A woman with systemic lupus erythematosus multiple pregnancy complications, and cerebral infarction who was only positive for phosphatidylserine-dependent antiprothrombin antibodies. Modern Rheumatology Case Reports. 2021;**5**:47-51. DOI: 10.1080/24725625.2020.1782030

[34] Žigon P, Pirkmajer KP, Tomšič M, Kveder T, Božič B, Šemrl SS, et al. Anti-phosphatidylserine/prothrombin antibodies are associated with adverse pregnancy outcomes. Journal of Immunology Research. 2015;**2015**:975704. DOI: 10.1155/2015/975704

[35] Hoxha A, Ruffatti A, Tonello M, Bontadi A, Salvan E, Banzato A. Antiphosphatidylserine/prothrombin antibodies in primary antiphospholipid syndrome. Lupus. 2012;**21**:787-789. DOI: 10.1177/0961203312441983

[36] Sciascia S, Sanna G, Murru V, Roccatello D, Khamashta M, Bertolaccini ML. Anti-prothrombin

(aPT) and anti-phosphatidylserine/prothrombin (aPS/PT) antibodies and the risk of thrombosis in the antiphospholipid syndrome:a systematic review. Thrombosis and Haemostasis. 2014;**111**:354-364. DOI: 10.1160/TH13-06-0509

[37] Chighizola CB, Raschi E, Banzato A, Borghi MO, Pengo V, Meroni PL. The challenges of lupus anticoagulants. Expert Review of Hematology. 2016;**9**:389-400. DOI: 10.1586/17474086.2016.1140034

[38] Andreoli L, Chighizola CB, Nalli C, Gerosa M, Borghi MO, Pregnolato F, et al. Clinical characterization of antiphospholipid syndrome by detection of IgG antibodies against β2-glycoprotein I domain 1 and domain 4/5: Ratio of anti-domain 1 to anti-domain 4/5 as a useful new biomarker for antiphospholipid syndrome. Arthritis & Rhematology. 2015;**67**:2196-2204. DOI: 10.1002/art.39187

[39] Sciascia S, Sanna G, Murru V, Roccatello D, Khamashta MA, Bertolaccini ML. The global anti-phospholipid syndrome score in primary APS. Rheumatology. 2015;**54**:134-138. DOI: 10.1093/rheumatology/keu307

[40] Sciascia S, Cuadrado MJ, Sanna G, Murru V, Roccatello D, Khamashta MA, et al. Thrombosis risk assessment in systemic lupus erythematosus: Validation of the global antiphospholipid syndrome in a prospective cohort. Arthritis Care and Research. 2014;**66**: 1915-1920. DOI: 10.1002/acr.22388

[41] Radin M, Cecchi I, Schreiber K, Rubini E, Roccatello D, Cuadrado MJ, et al. Pregnancy success rate and response to heparins and/or aspirin differ in women with antiphospholipid antibodies according to their global antiphospholipid syndrome score. Seminars in Arthritis and Rheumatism. 2020;**50**:553-556. DOI: 10.1016/j.semarthrit.2020.01.007

[42] de Groot PG, Urbanus RT. Antiphospholipid syndrome not a non-inflammatory disease. Seminars in Thrombosis and Hemostasis. 2015;**41**:607e14. DOI: 10.1055/s-0035-1556725

[43] Evain-Brion D, Guibourdenche J, Tsatsaris V, Fournier T. Human trophoblast differentiation. Bulletin de l'Academie Nationale de Medecine. 2009;**193**:1017-1025

[44] Mahler M, Norman GL, Meroni PL, Khamashta M. Autoantibodies to domain 1 of beta 2 glycoprotein 1: A promising candidate biomarker for risk management in antiphospholipid syndrome. Autoimmunity Reviews. 2012;**12**:313-317. DOI: 10.1016/j.autrev.2012.05.006

[45] Meroni PL, Di Simone N, Testoni C, D'Asta M, Acaia B, Caruso A. Antiphospholipid antibodies as cause of pregnancy loss. Lupus. 2004;**13**:649-652. DOI: 10.1191/0961203304lu2001oa

[46] Meroni PL, Borghi MO, Grossi C, Chighizola CB, Durigutto P, Tedesco F. Obstetric and vascular antiphospholipidi syndrome: Same antibodies but different diseases? Nature Reviews Rheumatology. 2018;**14**:433-440. DOI: 10.1038/s41584-018-0032-6

[47] Harper BE, Willis R, Pierangeli SS. Pathophysiology meccanism in antiphospholipid syndrome. International Journal of Clinica Rheumatology. 2011;**6**:157-171. DOI: 10.2217/ijr.11.9

[48] De Wolf F, Carreras LO, Moerman P, Vermylen J, Van Assche Am Renaer M. Decidual vasculopathy and extensive placental infarction in a patient with repeated thromboembolic accidents, recurrent fetal loss and a lupus anticoagulant. American Journal of Obstetrics and Gynecology. 1982;**142**:829-834. DOI: 10.1016/s0002-9378(16)32527-3

[49] Tong M, Viall CA, Chamley LW. Antiphospholipid antibodies and the placenta: A systematic review of their in vitro effects and modulation by treatment. Human Reproduction. 2015;**21**:97.118. DOI: 10.1093/humupd/dmu049

[50] Lockshin MD, Druzin ML, Goei S, Qamar T, Magin MS, Jovanovic L, et al. Antibody to cardiolipin as a predictor of fetal distress or death in pregnant patients with systemic lupus erythematosus. New England Journal of Medicine. 1985;**313**:152-156. DOI: 10.1056/NEJM198507183130304

[51] Sebire NJ, Fox H, Backos M, Rai R, Paterson C, Regan L. Defective endovascular trophoblast invasion in primary antiphospholipid antibody syndrome-associated early pregnancy failure. Human Reproduction. 2002;**17**:1067-1071. DOI: 10.1093/humrep/17.4.1067

[52] Viall CA, Chamley LW. Histopathology in the placentae of women with antiphospholipid antibodies: A systematic review of the literature. Autoimmunity Reviews. 2015;**14**:446-471. DOI: 10.1016/j.autrev.2015.01.008

[53] Marchetti T, Cohen M, de Moerloose P. Obstetrical antiphospholipid syndrome: From the pathogenesis to the clinical and therapeutic implications. Clinical and Development Immunology. 2013;**2013**:159124. DOI: 10.1155/2013/159124

[54] Simone D, Castellani R, Raschi E, Borghi MO, Meroni PL, Caruso A, et al. Anti-beta-2glycoprotein I antibodies affect Bcl-2 and Bax trophopblast expression without evidence of apoptosis. Annals of New York Academy of Sciences. 2006;**1069**:462-467. DOI: 10.1196/annals.1351.034

[55] Chen Q, Viall C, Kang Y, Liu B, Stone P, Chamley L. Anti-phopholipid antibodies increase non-apoptotic trophoblast shedding: A contribution to the pathogenesis of pre-eclampsia in affected women? Placenta. 2009;**30**:767-773. DOI: 10.1016/j.placenta.2009.06.008

[56] Marchetti T, Ruffatti A, Wuillemin C, de Moerloose P, Cohen M. Hydroxychloroquine restores trophoblast fusion affected by antiphospholipid antibodies. Journal of Thrombosis and Haemostasis. 2014;**12**:910-920. DOI: 10.1111/jth.12570

[57] Mulla MJ, Brosens JJ, Chamley LW, Giles I, Pericleous C, Rahman A, et al. Antiphospholipid antibodies induce a pro-inflammatory response in first trimester trophoblast via the TLR4/MyD88 pathway. American Journal of Reproduction Immunology. 2009;**62**:96-111. DOI: 10.1111/j.1600-0897.2009.00717.x

[58] Baumann MU, Bersinger NA, Mohaupt MG, Raio L, Gerber S, Surbek DV. First-trimester serum levels of soluble endoglin and soluble fms-like tyrosine kinase-1 as first-trimester markers for late-onset preeclampsia. American Journal of Obstetrics and Gynecology. 2008;**199**:266.e1-266.e6. DOI: 10.1016/j.ajog.2008.06.069

[59] Girardi G, Redecha P, Salmon JE. Heparin prevents antiphospholipid antibody-induced fetal loss by inhibiting complement activation. Nature Medicine. 2004;**10**:1222-1226. DOI: 10.1038/nm1121

[60] Salmon JE, Girardi G, Holer VM. Complement activation as a mediator of antiphospholipid antibody induced pregnancy loss and thrombosis. Annals of Rheumatic Diseases. 2002;**61**:ii46-ii50. DOI: 10.1136/ard.61.suppl_2.ii46

[61] Agostinis C, Durigutto P, Sblattero D, Borghi MO, Grossi C, Guida F, et al. A non-complement-fixing antibody to β2 glycoprotein I as a novel therapy for antiphospholipid

syndrome. Blood. 2014;**123**:3478-3487. DOI: 10.1182/blood-2013-11-537704

[62] Shamonki JM, Salmon JE, Hyjek E, Baergen RN. Excessive complement activation is associated with placental injury in patients with antiphospholipid antibodies. American Journal of Obstetrics and Gynecology. 2007;**196**: 167e161-167e165. DOI: 10.1016/j.ajog.2006.10.879

[63] Kim MY, Guerra MM, Kaplowitz E, Laskin CA, Petri M, Branch DW, et al. Complement activation predicts adverse pregnancy outcome in patients with systemic lupus erythematosus and/or antiphospholipid antibodies. Annals of Rheumatic Diseases. 2018;**77**:549-555. DOI: 10.1136/annrheumdis-2017-212224

[64] Berman J, Girardi G, Salmon JE. TNF-α is a critical effector and a target for therapy in antiphospholipid-induced pregnancy loss. Journal of Immunology. 2005;**174**:485-490. DOI: 10.4049/jimmunol.174.1.485

[65] Rand JH, Wu XX, Quin AS, Ashton AW, Chen PP, Hathcock JJ, et al. Hydroxychloroquine protects the annexin A5 anticoagulant shield from disruption by antiphospholipid antibodies: Evidence for a novel effect for an old antimalarial drug. Blood. 2010;**115**:2292-2299. DOI: 10.1182/blood-2009-04-213520

[66] Irman S, Škarabot M, Muševič I, Rozman B, Božič B. Thrombomodulatory effect of anti-B2-glycoprotein I antibodies on crystalline annexin A5 on phospholipid bilayers, as observed by atomic force microscopy. EJIFCC. 2011;**3**:81-93

[67] Alijotas-Reig J, Esteve-Valverde E, Ferrer-Oliveras R, Sáez-Comet L, Lefkou E, Mekinian A, et al. The European registry on obstetric antiphospholipid syndrome (EUROAPS): A survey of 1000 consecutive cases. Autoimmunity Reviews. 2019;**18**:406-414. DOI: 10.1016/j.autrev.2018.12.006

[68] Abou-Nassar K, Carrier M, Ramsay T, Rodger MA. The association between antiphospholipid antibodies and placenta mediated complications: A systematic review and meta-analysis. Thrombosis Research. 2011;**128**:77-85. DOI: 10.1016/j.thromres.2011.02.006

[69] Silver RM, Parker CB, Reddy UM, Goldenberg R, Coustan D, Dudley DJ, et al. Antiphospholipid antibodies in stillbirth. Obstetrics & Gynecology. 2013;**122**:641-657. DOI: 10.1097/AOG.0b013e3182a1060e

[70] Clark EA, Silver RM, Branch DW. Do antiphospholipid antibodies cause preeclampsia and HELLP syndrome? Current Rheumatology Reports. 2007;**9**:219-225. DOI: 10.1007/s11926-007-0035-9

[71] Gibbins KJ, Tebo A, Nielsen SK, Branch DW. Antiphospholipid antibodies in women with severe pre-eclampsia and placental insufficiency: A case-control study. Lupus. 2018;**27**:1903-1910. DOI: 10.1177/0961203318787035

[72] Saccone G, Berghella V, Maruotti GM, Ghi T, Rizzo G, Simonazzi G, et al. Antiphospholipid antibody profile based obstetric outcomes of primary antiphospholipid syndrome: The PREGNANTS study. American Journal of Obstetrics and Gynecology. 2017;**216**:525:e1-525 e12. DOI: 10.1016/j.ajog.2017.01.026

[73] Ruffatti A, Tonello M, Visentin MS, Bontadi A, Hoxha A, De Carolis S, et al. Risk factors for pregnancy failure in patients with antiphospholipid syndrome treated with conventional therapies: A multicentre, case-control study. Rheumatology (Oxford). 2011;**50**:1684-1689. DOI: 10.1093/rheumatology/ker139

[74] Sailer T, Zoghlami C, Kurz C, Rumpold H, Quehenberger P, Panzer S, et al. Anti-beta2-glycoprotein I antibodies are associated with pregnancy loss in women with the lupus anticoagulant. Thrombosis and Haemostasis. 2006;**95**:796-801

[75] Lockshin MD, Kim M, Laskin CA, Guerra M, Branch DW, Merrill J, et al. Prediction of adverse pregnancy outcome by the presence of lupus anticoagulant, but not anticardiolipin antibody, in patients with antiphospholipid antibodies. Arthritis & Rheumatology. 2012;**64**:2311-2318. DOI: 10.1002/art.34402

[76] Matsuki Y, Atsumi T, Yamaguchi K, Hisano M, Arata N, Oku K, et al. Clinical features and pregnancy outcome in antiphospholipid syndrome with history of severe pregnancy complications. Modern Rheumatology. 2015;**25**:215-218. DOI: 10.3109/14397595.2014.942503

[77] Latino JO, Udry S, Aranda FM, Peres Wingeyer SDA, Fernandez Romero DS, de Larranaga GF. Pregnancy failure in patients with obstetric antiphospholipid syndrome with conventional treatment: The influence of a triple positive antibody profile. Lupus. 2017;**26**:983-988. DOI: 10.1177/0961203317692432

[78] Ruffatti A, Calligaro A, Hoxha A, Trevisanuto D, Ruffatti AT, Gervasi MT, et al. Laboratory and clinical features of pregnant women with antiphospholipid syndrome and neonatal outcome. Arthritis Care and Research (Hoboken). 2010;**62**:302-307. DOI: 10.1002/acr.20098

[79] Bouvier S, Cochery-Nouvellon E, Lavigne-Lissalde G, et al. Comparative incidence of pregnancy outcome in treated obstetric antiphospholipid syndrome: The NOH-APS observational study. Blood. 2014;**123**:404-414. DOI: 10.1182/blood-2013-08-522623

[80] Hoxha A, Favaro M, Calligaro A, Del Ross T, Ruffatti AT, Infantolino C, et al. Upgrading therapy strategy improves pregnancy outcome in antiphospholipid syndrome: A cohort management study. Thrombosis and Haemostasis. 2020;**120**:36-43. DOI: 10.1055/s-0039-1697665

[81] Le Thi Huong DW, Wechsler B, Vauthier-Brouzes D, Duhaut P, Costedoat N, Andreu MR, et al. The second trimester Doppler ultrasound examination is the best predictor of late pregnancy outcome in systemic lupus erythematosus and/or the antiphospholipid syndrome. Rheumatology. 2006;**45**:332-338. DOI: 10.1093/rheumatology/kei159

[82] Andreoli L, Bertsia GK, Agmon-Levi N, Brown S, Cervera R, Costedoat-Chalumeau N, et al. EULAR recommendation for women's health and the management of family planning, assisted reproduction, pregnancy and menopause in patients with systemic lupus erythematosus and/or antiphospholipid syndrome. Annals of Rheumatic Diseases. 2017;**76**:476-485. DOI: 10.1136/annrheumdis-2016-209770

[83] Tektonidou MG, Andreoli L, Limper M, Amoura Z, Cervera R, Costedoat-Chalumeau N, et al. EULAR recommendations for the management of antiphospholipid syndrome in adults. Annals of Rheumatic Disease. 2019;**78**:1296-1304. DOI: 10.1136/annrheumdis-2019-215213

[84] Kutteh WH. Antiphospholipid antibody-associated recurrent pregnancy loss: Treatment with heparin and low-dose aspirin is superior to low-dose aspirin alone. American Journal of Obstetrics & Gynecology. 1996;**174**:1584e9. DOI: 10.1016/s0002-9378(96)70610-5

[85] Rai R, Cohen H, Dave M, Regan L. Randomized controlled trial of aspirin and aspirin plus heparin in pregnant

women with recurrent miscarriage associated with phospholipid antibodies (or antiphospholipid antibodies). BMJ. 1997;**314**:253e7. DOI: 10.1136/bmj.314.7076.253

[86] Farquharson RG, Quenby S, Greaves M. Antiphospholipid syndrome in pregnancy: A randomized, controlled trial of treatment. Obstetrics and Gynecology. 2002;**100**:408e13. DOI: 10.1016/s0029-7844(02)02165-8

[87] Laskin CA, Spitzer KA, Clark CA, Crowther MR, Ginsberg JS, Hawker GA, et al. Low molecular weight heparin and aspirin for recurrent pregnancy loss: Results from the randomized, controlled HepASA trial. Journal of Rheumatology. 2009;**36**:279e87. DOI: 10.3899/jrheum.080763), 10.3899/jrheum.080763)

[88] Mak A, Cheung MW, Cheak AA, Ho RC. Combination of heparin and aspirin is superior to aspirin alone in enhancing live births in patients with recurrent pregnancy loss and positive anti-phospholipid antibodies: A meta-analysis of randomized controlled trials and meta-regression. Rheumatology (Oxford). 2010;**49**:281-288. DOI: 10.1093/rheumatology/kep373

[89] Hamulyák EN, Scheres LJ, Marijnen MC, Goddijn M, Middeldorp S. Aspirin or heparin or both for improving pregnancy outcomes in women with persistent antiphospholipid antibodies and recurrent pregnancy loss. Cochrane Database of Systematic Reviews. 2020;**5**:CD012852. DOI: 10.1002/14651858.CD012852

[90] Ruffatti A, Gervasi MT, Favaro M, Ruffatti AT, Hoxha A, Punzi L. Adjusted prophylactic doses of nadroparin plus low dose aspirin therapy in obstetric antiphospholipid syndrome. A prospective cohort management study. Clinical and Experimental and Rheumatology. 2011;**29**:551-554

[91] Arachchillage DR, Machin SJ, Mackie IJ, Cohen H. Diagnosis and management of non-criteria obstetric antiphospholipid syndrome. Thrombosis and Haemostasis. 2015;**113**:13-19. DOI: 10.1160/TH14-05-0416

[92] Lassere M, Empson M. Treatment of antiphospholipid syndrome in pregnancy—A systematic review of randomized therapeutic trials. Thrombosis Research. 2004;**114**: 419-426. DOI: 10.1016/j.thromres.2004.08.006

[93] Ruffatti A, Favaro M, Calligaro A, Zambon A, Del Ross T. Management of pregnant women with antiphospholipid antibodies. Expert Review in Clinical Immunology. 2019;**15**:347-358. DOI: 10.1080/1744666X.2019.1565995

[94] Bramham K, Thomas M, Nelson-Piercy C, Khamashta M, Hunt BJ. First-trimester low-dose prednisolone in refractory antiphospholipid antibody-related pregnancy loss. Blood. 2011;**117**:6948-6951. DOI: 10.1182/blood-2011-02-339234

[95] Ruffatti A, Salvan E, Del Ross T, Gerosa M, Andreoli L, Maina A, et al. Treatment strategies and pregnancy outcomes in antiphospholipid syndrome patients with thrombosis and triple antiphospholipid positivity. a European multicentre retrospective study. Thrombosis and Haemostasis. 2014;**112**:727-735. DOI: 10.1160/TH14-03-0191

[96] Mekinian A, Lazzaroni MG, Kuzenko A, Alijotas-Reig J, Ruffatti A, Levy P, et al. The efficacy of hydroxychloroquine for obstetrical outcome in anti-phospholipid syndrome: Data from a European multicenter retrospective study. Autoimmunity Reviews. 2015;**14**:498-502. DOI: 10.1016/j.autrev.2015.01.012

[97] El-Haieg DO, Zanati MF, El-Foual FM. Plasmapheresis and

pregnancy outcome in patients with antiphospholipid syndrome. International Journal of Gynaecology and Obstetrics. 2007;**99**:236-241. DOI: 10.1016/j.ijgo.2007.05.045

[98] Heilmann L, Schorch M, Hahn T, Adasz G, Schilberz K, Adiguzel C, et al. Pregnancy outcome in women with antiphospholipid antibodies: Report on a retrospective study. Seminars in Thrombosis and Hemostasis. 2008;**34**: 794-802. DOI: 10.1055/ s-0029-1145261

[99] Ruffatti A, Tonello M, Hoxha A, Sciascia S, Cuadrado MJ, Latino JO, et al. Effect of additional treatments combined with conventional therapies in pregnant patients with high-risk antiphospholipid syndrome: A multicentre study. Thrombosis and Haemostasis. 2018;**118**:639-646. DOI: 10.1055/s-0038-1632388

[100] Stojanovich L, Mikovic Z, Mandic V, Popovich-Kuzmanovich D. Treatment of anti-phospholipid syndrome in pregnancy with low doses of intravenous immunoglobulin. Israel Medical Association Journal. 2007;**9**:555-556

[101] Chang P, Millar D, Tsang P, Lim K, Houlihan E, Stephenson M. Intravenous immunoglobulin in antiphospholipid syndrome and maternal floor infarction when standard treatment fails: A case report. American Journal of Perinatology. 2006;**23**:125-129. DOI: 10.1055/s-2006-931805

[102] Watanabe N, Yamaguchi K, Motomura K, Hisano M, Sago H, Murashima A. Combination therapy with anticoagulants, corticosteroids and intravenous immunoglobulin for women with severe obstetric antiphospholipid syndrome. Clinical and Experimental Rheumatology. 2014;**32**:299-300

[103] De Carolis S, Botta A, Salvi S, di Pasquo E, Del Sordo G, Garufi C, et al. Is there any role for the hydroxychloroquine (HCQ) in refractory obstetrical antiphospholipid syndrome (APS) treatment? Autoimmunity Reviews. 2015;**14**:760-762. DOI: 10.1016/j. autrev.2015.04.010

[104] Mayer-Pickel K, Horn S, Lang U, Cervar-Zivkovic M. Response to plasmapheresis measured by a ngiogenic factors in a woman with antiphospholipid syndrome in pregnancy. Case Reports in Obstetrics and Gynecology. 2015;**2015**:123408. DOI: 10.1155/2015/ 123408

[105] Abrahams VM, Chamley LW, Salmon JE. Antiphospholipid syndrome and pregnancy: Pathogenesis to translation. Arthritis & Rheumatology. 2017;**69**:1710e4. DOI: 10.1002/art.40136

[106] Lefkou E, Mamopoulos A, Dagklis T, Vosnakis C, Rousso D, Giradi G, et al. Pravastatin improves pregnancy outcomes in obstetric antiphospholipid syndrome refractory to antithrombotic therapy. The Journal of Clinical Investigation. 2016;**126**: 2933e40. DOI: 10.1172/JCI86957

[107] Costantine MM, Cleary K. Pravastatin for the prevention of preeclampsia in high-risk pregnant women. Obstetrics & Gynecology. 2013;**121**:349e53. DOI: 10.1097/ AOG.0b013e31827d8ad5

[108] Belizna C, Pregnolato F, Abad S, Alijotas-Reig J, Amital H, Amoura Z, et al. HIBISCUS: Hydroxychloroquine for the secondary prevention of thrombotic and obstetrical events in primary antiphospholipid syndrome. Autoimmunity Reviews. 2018;**17**:1153-1168. DOI: 10.1016/j.autrev.2018.05.012

[109] Schreiber K, Breen K, Cohen H, Jacobsen S, Middeldorp S, Pavord S, et al. Hydroxychloroquine to improve pregnancy outcome in women with antiphospholipid antibodies (HYPATIA) protocol: A multinational randomized controlled trial of

hydroxychloroquine versus placebo in addition to standard treatment in pregnant women with antiphospholipid syndrome or antibodies. Seminars in Thrombosis and Hemostasis. 2017;**43**:562-571. DOI: 10.1055/s-0037-1603359

Chapter 4

Bleeding in Patients with Antiphospholipid Antibodies

Peter Kubisz, Pavol Holly and Jan Stasko

Abstract

The antiphospholipid antibodies (aPL) are commonly associated with thrombotic events and obstetric complications. However, apart from the bleeding complications of antithrombotic therapy, the acquired coagulopathy caused by the aPL, particularly by lupus anticoagulant and anticardiolipin antibodies, might be occasionally manifested as a hemorrhagic syndrome with various clinical severity. Bleeding symptoms vary from mild (mucocutaneous) up to life-threatening (gastrointestinal, intracranial). The bleeding may be the first manifestation of aPL or appear concomitantly with thrombosis. The underlying hemostatic changes include thrombocytopenia, platelet function disorders, and coagulation factor inhibitors or deficiencies, namely prothrombin, FVII, FVIII, FX, and FXI. Thrombocytopenia is the most common finding, seen in up to 53% of patients with aPL, although it is usually mild to moderate and associated with significant bleeding only in a minority of cases. Of interest, patients with severe thrombocytopenia appear to be less likely to suffer from thrombotic events. The involved pathophysiological mechanisms are heterogeneous. Non-neutralizing antibodies against coagulation factors resulting in increased clearance, specific antibodies against platelet membrane glycoproteins, increasing platelet activation and aggregation with subsequent consumption, and immune-mediated platelet clearance are among those identified. Immunosuppression, preferably with corticosteroids, represents the first-choice therapeutic approach. Plasmapheresis is efficient in the case of catastrophic antiphospholipid syndrome. Antithrombotic therapy can be challenging, but its administration should continue as much as possible.

Keywords: hemorrhage, antiphospholipid antibodies, thrombocytopenia, acquired prothrombin deficiency, acquired coagulation factor deficiencies, coagulation factor inhibitors

1. Introduction

The antiphospholipid syndrome (APS) is an acquired autoimmune disorder, defined by the combination of generally accepted laboratory and clinical criteria [1]. The latest laboratory criteria include repeated (at least 12 weeks apart) positive testing for at least 1 of 3 selected antiphospholipid antibodies (aPL): lupus anticoagulant (LA), anticardiolipin (aCL), anti-beta2-glycoprotein I (anti-B2GPI) antibodies. Clinical criteria emphasize the arterial and venous thromboembolic and pregnancy-related (recurrent miscarriages in the first trimester, fetal death in the second or third trimesters, severe pre-eclampsia requiring delivery of a premature infant before 34 weeks of gestation) events. However, other laboratory

IntechOpen

and clinical complications with clear association to aPL, referred to as non-criteria manifestations, have been described. Based on the affected organ system, the clinical non-criteria manifestations divide into several subgroups: cardiovascular, neurologic, skin, renal, hematologic, and other [2, 3]. Hematologic complications include thrombocytopenia, hemolytic anemia, and functional changes or deficiencies of coagulation factors with both thrombotic (acquired resistance to activated protein C, protein S deficiency) or bleeding tendencies. As mentioned above, the association of aPL with thromboembolic events is extensively and well documented. However, the acquired coagulopathy caused by the aPL is complex and

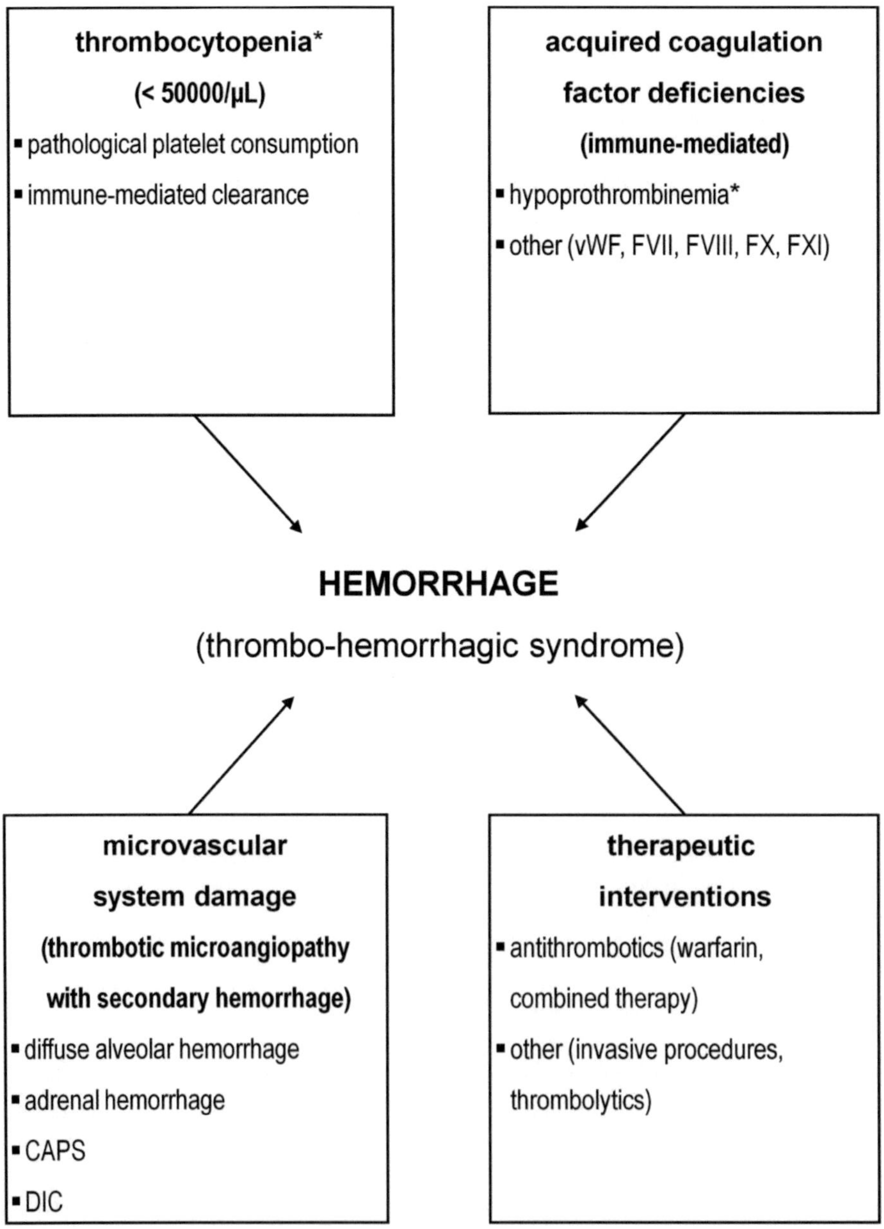

Figure 1.
*Pathomechanisms involved in hemorrhage in aPL patients. CAPS, catastrophic antiphospholipid syndrome; DIC, disseminated intravascular coagulation; F, factor; vWF, von Willebrand factor; * most common pathomechanisms.*

might occasionally manifest as a hemorrhagic event with various clinical severity or combined thrombo-hemorrhagic syndrome. The latter is common in catastrophic APS (CAPS), a rare but often fatal variant with excessive activation of hemostasis, consumption of its components, and micro-thrombotic damage in multiple organs.

aPL can interact with different blood and vascular components and cause hemorrhage through several mechanisms (**Figure 1**) [1]. Firstly, aPL-positive patients frequently develop thrombocytopenia. Secondly, acquired immune-mediated coagulation factor deficiencies, such as hypoprothrombinemia, can appear after the interaction between aPL and coagulation factors.

Thirdly, the microvascular system damage with an extensive thrombotic or inflammatory insult via the monocyte, endothelial, and complement activation can result in secondary bleeding to the affected tissue. Thrombotic microangiopathies (TMAs) such as CAPS, as well as diffuse alveolar hemorrhage (DAH) and adrenal hemorrhage (AH), the pathognomic complications of APS, are representative examples of this pathomechanism. Since the antithrombotic therapy remains a mainstay of management of aPL, the extensive use of antithrombotics, typical for patients afflicted with their presence, can contribute to bleeding events and represents the fourth cause. Severe thrombocytopenia (platelet count lower than 50000/μL) and prothrombin deficiency are the most prominent causes of bleeding [4]. The discussion of the given pathomechanisms follows.

2. Thrombocytopenia in patients with aPL

Though not included in the current diagnostic criteria for APS (Sydney 2012 criteria), thrombocytopenia represents a complication directly linked to aPL [1]. Thrombocytopenia is a frequent finding in aPL-positive patients; it is their most common non-criteria hematologic manifestation. The Euro-Phospholipid project, a large prospective multicenter international study evaluating 1000 European patients with both primary and secondary APS, found thrombocytopenia in 296 (29.6%) of its participants [5]. Other studies focused on the whole population of aPL-positive patients reported comparable incidence, ranging from 20 to 53% [6–10]. Of interest, particular subgroups seem to be more prone to develop thrombocytopenia. Patients with secondary APS associated with systemic lupus erythematosus (SLE) have thrombocytopenia approximately 2-times more often than those with primary APS (reported incidence 40 vs. 21% in the Euro-Phospholipid project) [5]. A low platelet count is more frequent in patients with CAPS [10, 11].

Thrombocytopenia tends to be mild to moderate with the nadir above 50000/μL in most cases. Only a small portion of patients (approximately 10%) develop severe thrombocytopenia, and its occurrence is often associated with other complications, such as TMAs (disseminated intravascular coagulation (DIC), CAPS) [8]. Rapid (within days) progression of thrombocytopenia or its new occurrence in patients with previously normal platelet count can be the first indication of CAPS [11, 12].

Despite being common, thrombocytopenia alone is not usually responsible for clinically relevant bleeding. For example, in the Italian Registry of aPL, only four patients out of 90 with thrombocytopenia experienced major hemorrhagic events [8]. On the other hand, nor it protects, especially if mild to moderate, from thromboembolism. Notwithstanding, if severe and without CAPS, it can account for a minor protective effect. In the Italian Registry with 360 patients included, severe thrombocytopenia was associated with a significantly lower rate of thrombotic events in comparison to the group with normal platelet count; however, the group with mild thrombocytopenia did not show a significant difference (9 vs. 40 vs. 32%) [8]. A recent study analyzing altogether 305 patients with primary

APS, 51 with thrombocytopenia included, observed a higher rate of thrombotic relapses (29% vs. 19%) during a long (median 11 years) follow-up in the group with thrombocytopenia, though the difference did not reach statistical significance [13]. Despite comparable antithrombotic therapy, no difference in major hemorrhage (4% vs. 3%) was observed between the thrombocytopenic and non-thrombocytopenic group, albeit the significantly higher rate of overall bleeding (17% vs. 8%) was in the thrombocytopenic group. The authors conclude that thrombocytopenia may have a prognostic value in primary APS and help identify high-risk patients for APS-related complications [13].

The evidence concerning the association between thrombocytopenia and other clinical features of APS such as hemolytic anemia, livedo reticularis, skin ulcerations, chorea, and cardiac valve dysfunction is conflicting. Some studies, but not all, observed more frequent occurrence of those symptoms in patients with thrombocytopenia.

The pathogenesis of aPL-related thrombocytopenia is likely heterogeneous. aPL can directly or indirectly via B2GPI interact with several platelet membrane glycoproteins (GP) and phospholipids and thus initiate two processes: 1) pathologically enhanced platelet activation and aggregation after their initial activation or damage with subsequent platelet thrombus formation and platelet consumption; 2) immune-mediated pathological platelet clearance. The interaction with platelets involves the binding of anti-B2GPI via B2GPI to the activated platelet surface or direct interaction of aPL with specific platelet membrane glycoproteins (GPIb/IX, GPIIb/IIIa, GPIV) [14]. Particular subtypes of aPL and their quantity likely play a prominent role in the pathogenesis of thrombocytopenia. Anti–B2GPI antibodies of IgG class, LA, a higher titer of aCL of IgG class, and triple aPL positivity were a more common finding in patients with thrombocytopenia [13, 15, 16]. LA and a high titer of aCL were frequent among patients with severe thrombocytopenia. Since LA is associated with the highest prothrombotic risk among aPL, its higher prevalence in these patients could mitigate the bleeding tendencies and contribute to a relatively low rate of major bleedings.

Other pathomechanisms may occasionally contribute to thrombocytopenia in aPL-positive patients. The association, albeit anecdotal, between aPL and the hemophagocytic syndrome (a hyperinflammatory disorder with pathological phagocytosis of blood cells and their precursors in the bone marrow and other tissues) and bone marrow necrosis was described [17, 18]. These disorders decrease platelets via impairing megakaryopoiesis. Splenomegaly after splenic or portal vein thrombosis leads to increased platelet pooling and redistribution from circulation [7].

It should be emphasized that the etiology of thrombocytopenia in aPL-positive patients can be multifactorial and not exclusively linked to these antibodies. Other diseases can contribute to and further deepen the decrease in platelet count. Coincidence with immune thrombocytopenia (ITP), drug-induced thrombocytopenia with heparin-induced thrombocytopenia included, thrombocytopenia related to infections, TMAs, and pregnancy-related thrombocytopenia have been described [18, 19].

The relationship with ITP seems to be particularly interesting and complex. Patients with ITP are frequently positively tested for aPL, with a reported incidence ranging from 25 to 75% [20]. A recent retrospective study of 159 adult patients with primary and secondary severe ITP (platelet count below 50000/μL) identified aPL in 37 (23.2%), with 14 being triple positive. Triple positivity was associated with a lower platelet count [21]. Clinical implications of the relation between ITP and aPL are still discussed and not clear. The available data regarding the risk of thrombosis and treatment are inconclusive. However, a recent study with altogether 196 patients with primary ITP, 49 aPL-positive included, did observe a significantly higher risk of thrombotic events. Other monitored characteristics (hemorrhage, response to therapy, clinical course, changes in platelet counts) were comparable [19]. Analogically to the observation in APS, it seems that the risk of thrombosis in patients with concomitant ITP and aPL positivity, particularly in those undergoing therapy with corticosteroids and other immunosuppressive agents, is more prominent than the risk of bleeding.

The diagnostic approach has to consider the possibility of aPL as a sole cause of thrombocytopenia as well as the coincidence of other disorders with aPL, especially TMAs. Since patients with aPL/APS are often anticoagulated and treated with immunosuppressives, heparin-induced thrombocytopenia and infectious causes should be addressed in the diagnostic process.

Since thrombocytopenia in aPL-positive patients is predominantly mild and without significant bleeding, outside of CAPS, most patients do not require specific treatment. As a general rule, the goal is to maintain the platelet count above 30000/μL – a critical threshold for the development of severe spontaneous bleeding. When immune etiology is behind thrombocytopenia, strategies effective in ITP are preferably used [22, 23]. Corticosteroids, initially in high-dose with gradual tapering, alone or combined with intravenous immunoglobulins (IVIg), represent the first-line treatment. In contrast to ITP, the use of IVIg as a first-line treatment is controversial in aPL-positive patients since their administration is potentially associated with an increased risk of thrombotic events [24]. Other immunosuppressive or immunomodulatory agents or procedures (danazol, chloroquine, dapsone, rituximab, plasmapheresis) or splenectomy can be chosen as alternatives for those with inadequate response. Rituximab seems to be a particularly perspective agent. Though only limited clinical data from a small number of patients are available so far, the response and persisting stable platelet count after rituximab have been observed in a reasonably high number (50–83%) of treated patients [25, 26]. It is important to emphasize that most of the included patients had refractory thrombocytopenia without a satisfactory response to previous treatment modalities. Rituximab was tolerated well with no significant increase in thrombotic risk. Its risk profile in the aPL setting appears to be comparable to ITP [25].

The use of thrombopoietin mimetics (TPOMs) remains controversial due to the conflicting clinical data. There is a general agreement on their effectiveness in increasing platelet count, but safety remains an open issue. Several authors did not observe any increase in the thrombotic events during the administration of TPOMs [27, 28]. Others, including those who analyzed larger patient groups, report a prothrombotic risk associated with this therapy in the a-PL positive subgroup [29–31]. Gonzales et al. found in their retrospective study of 46 patients with thrombocytopenia and various systemic autoimmune disorders, all treated with eltrombopag, that 3 (6.5%) participants suffered from thrombotic events while on treatment. Crucially, 6 out of 46 participants had concurrent APS, and 2 of them (33% of all patients with aPL) were among those with thrombosis [30]. Guitton et al. retrospectively studied 18 patients with thrombocytopenia and SLE treated with romiplostim or eltrombopag; 10 had been diagnosed with concurrent aPL/APS. 5 patients developed thrombosis; 3 of them (30% of all patients with aPL) had APS [31]. These observations suggest a higher thrombotic risk in the aPL-positive group. Though well established in therapy of ITP, the use of TPOMs in aPL-positive patients requires caution and individual evaluation of thrombotic risk. Minimized dosing of TPOMs, aimed to maintain platelet count around 50000/μL, was suggested to decrease thrombotic risk since the thrombotic events are more frequent at platelet count greater than 100000/μL [22].

Except for severe thrombocytopenia, a decrease in platelet count does not fully protect patients with aPL/APS from thromboembolic events, and antithrombotic prevention or therapy should be continuing as long as possible. However, bleeding risk has to be considered, and an individualized approach is mandatory. In general, full anticoagulation can be given in the setting of platelet count over 50000/μL, and its stopping should be considered seriously in platelet count below 25000/μL. The patients with platelets between these values should be treated individually with anticoagulants in reduced doses. Half-dose low molecular weight heparins (LMWHs) represent the usual first-choice treatment [22].

3. Factor deficiencies associated with aPL

3.1 Hypoprothrombinemia

Acquired deficiency of prothrombin, referred to as lupus anticoagulant hypoprothrombinemia syndrome, is the most known and well defined of all coagulation factor deficiencies associated with aPL. Its precise incidence is unclear, but with the order of magnitude of hundreds of reported cases, it appears to be a rare complication [32, 33]. It typically occurs in the child or adolescent female patients with aPL after viral infections or with systemic immune disorders, most commonly SLE [34]. Adults can be affected as well, albeit less frequently [35]. The preexisting systemic immune disease is not obligatory since cases without were identified; other precipitating conditions include tumors such as lymphomas, particularly with the production of pathological immunoglobulins and drug reactions.

Bleeding severity varies from mild mucocutaneous (epistaxis, ecchymosis), which is the most common, to severe and life-threatening, including localizations such as muscles, genitourinary tract, gastrointestinal tract (GIT), and central nervous system (CNS) [32–38]. A substantial number of patients (up to 50%) have no significant bleeding events and can be even asymptomatic [36]. Concomitant presence of thrombotic events, hemorrhagic-thrombotic syndrome, and CAPS were occasionally described [39–41]. The condition is usually self-limiting when associated with viral infections, whereas it can have a lasting duration or relapses in the presence of autoimmune diseases [36]. Despite the possibility of severe bleeding events, the overall prognosis is good in general, with a reported mortality rate of less than 5%.

Laboratory findings include the prolongation of both prothrombin (PT) and activated partial thromboplastin time (aPTT), variably decreased prothrombin activity (about 10–20% on average, although it may be extremely low or unmeasurable) with a proportional decrease of prothrombin antigen. As mentioned above, a deficiency of other coagulation factors might be present. Therefore, their activity should be checked [32]. Positive testing for LA complements the picture. The finding of PT prolongation in an aPL-positive patient should prompt the testing for prothrombin deficiency even if no bleeding is apparent at the time.

The traditional view based on the initial analyses in the 1980s defined the involved antibodies as non-neutralizing, unable to directly inhibit the prothrombin coagulation activity [42]. Cross-reactivity between the aPL and phospholipid epitopes in the prothrombin molecules is a likely explanation. The aPL form prothrombin antigen–antibody complexes, and their subsequent elimination results in the proportional decrease of both prothrombin activity and antigen. If the clearance is extensive enough to lead to a relevant prothrombin decrease with its activity below 20%, bleeding manifestations may occur. However, some researchers provided conflicting evidence with hints on more complex changes of hemostasis. In the recent analysis of a relatively large cohort of 41 patients, Japanese authors did not observe an exact correlation between prothrombin levels, anti-FII antibody quantity, and bleeding phenotype. They also identify different autoantibodies directed against FVIII besides the anti-prothrombin ones in several patients with the disorder [43]. They confirmed combined coagulation factor deficiencies in a small number of the studied cases as well. Based on this observation and a known heterogeneity of the clinical presentation, it is reasonable to conclude that hypoprothrombinemia is not an isolated change in aPL-positive patients, and a complex evaluation of hemostasis is always required.

The therapeutic approach aims at (1) stopping the active bleeding; (2) eradicating antibodies responsible for prothrombin deficiency; (3) preventing further thromboembolic events [35, 37]. The withdrawal of antithrombotic agents, supplementation of blood components (transfusion of packed red blood cells and fresh frozen plasma), activation of coagulation factor production (vitamin K administration), hemostatic agents (styptics, antifibrinolytics) represent the strategies for bleeding cessation [35]. However, all these approaches can in aPL-positive patients, especially in prolonged use, lead to the increased thrombotic risk. Immunosuppression, with corticosteroids as the first-line choice or other agents (azathioprine, rituximab, cyclophosphamide) and procedures (plasma exchange) as alternatives, leads to antibody eradication. Monotherapy with corticosteroids is efficient in most cases. Measurement of prothrombin levels, whether by clotting, chromogenic or immunologic methods, can be used for the treatment monitoring. Since the risk of thrombosis usually remains significantly increased even in the presence of bleeding and bleeding itself does not protect from thromboembolism, the therapies aimed at bleeding cessation has to be counterweighted by antithrombotic therapy. Both bleeding and thromboembolic risks have to be evaluated carefully in all cases.

3.2 von Willebrand factor deficiency (von Willebrand syndrome)

Few case reports of concurrent acquired von Willebrand syndrome (AWS), an acquired vWF deficiency, with the presence of aPL were described [44–48]. Interestingly, other disorders with well-defined relation to AWS (myeloproliferative neoplasm, aortic valve stenosis, connective tissue diseases such as SLE) were identified in most cases. Therefore, aPL are not regarded as a usual cause of AWS, but rather as a coincidental finding in underlying immune disorders. Some researchers speculated that aPL might modify and counterbalance the bleeding phenotype typical for AWS [44, 48]. Thrombotic event after normalization of vWF was reported [44]. Immunosuppression, the standard treatment of AWS, combined with antithrombotic prevention, was given in reported cases with good clinical outcomes.

3.3 Deficiencies of other coagulation factors

Acquired deficiencies of other clotting factors, namely FVII, FVIII, FX, and FXI, were reported [49–52]. In summary, these deficiencies are extremely rare, and clinical data are limited to case reports or series. Bleeding manifestations are variable, with varying severity. The therapeutic strategies are similar to the approach used in AWS.

4. Bleeding in thrombotic microangiopathies associated with aPL

4.1 Diffuse alveolar hemorrhage

DAH is a severe and life-threatening pulmonary complication of aPL. Inflammatory damage to the pulmonary microcirculation, namely to alveolar arterioles, capillaries, and venules, with subsequent necrotic changes and secondary hemorrhage, define the disorder [53]. A microscopic pathoanatomical picture typically reveals capillaritis with interstitial neutrophilic infiltrate, thrombi in small muscular pulmonary arteries, myointimal thickening, and the remodeling of the muscular pulmonary arteries and arterioles [53, 54]. The condition is rare and appears in less than 1% of all aPL-positive patients, though it is considerably more frequent and clinically relevant in those with CAPS, affecting 5–10% [54–57]. Both genders are affected, but males constitute approximately 2/3 of cases with primary APS, whereas women dominate the group with APS secondary to SLE [54]. The patients with DAH are more likely to have a higher titer of aPL and suffer from other comorbidities associated with aPL than those without DAH. Cardiac valve disease, pulmonary hypertension, livedo reticularis, skin ulcers, CNS involvement (stroke or seizure), and pregnancy complications are among the reported concomitant disorders [54, 57].

Several pathomechanisms are likely to participate in the damage of the alveolar structures in DAH in aPL-positive patients. aPL-mediated activation of endothelial cells, resulting in the increased expression of tissue factor, platelet, and complement activation with C5a-mediated neutrophil recruitment and the subsequent lung tissue injury is likely behind thrombi formation in the pulmonary microcirculation. aPL-induced systemic inflammatory response syndrome with the excessive cytokine activation (e. g. tumor necrosis factor-α, interleukin-1, interleukin-6, interleukin-18, macrophage migration inhibitory factor) as well as L-ficolin-induced lung injury and interstitial neutrophilic infiltration lead to the loss of the integrity of the alveolar-capillary basement membrane. Disruption of alveolar structure and its veins through the combination of inflammation and thrombosis result in the extravasation of red blood cells into the alveoli [58].

The usual clinical presentation of DAH includes fever, chest pain, cough with hemoptysis, and dyspnea with the signs of hypoxemic respiratory failure [4]. However, not all symptoms, including hemoptysis, need to be present in every case. The symptoms are not specific and appear in other pulmonary diseases such as pulmonary embolism, pneumonia, and pulmonary edema. The complex differential diagnostics is of utmost importance. Laboratory and complementary tests are critical for the distinction of DAH. Anemia, aPL positivity, high diffusing capacity for carbon monoxide in pulmonary function tests, patchy or perihilar opacities on the chest X-ray and signs of hemorrhage, ground-glass infiltrates, and reticular interstitial opacities on pulmonary CT scans belong to the typical findings. Bronchoscopy with bronchoalveolar lavage and biopsy can document alveolar hemorrhage, exclude infections, and provide biological material for cytologic and histologic evaluation. Lung biopsy remains the gold standard for the definitive diagnosis, albeit the patient's condition and benefit–risk ratio should be evaluated before the procedure. As mentioned above, DAH is quite frequently associated with CAPS. Treating physicians should actively search for its signs in all cases.

Immune suppression remains the mainstay of the therapy. High-dose corticosteroids are the preferred initial treatment. The use of other immunosuppressives remains without a clear consensus due to the rarity of the condition and limited clinical data. However, available clinical data support the combined immune suppression (corticosteroids plus another immunosuppressive agent) over monotherapy with corticosteroids. The combined therapy seems to improve the clinical outcome and rate of long-term remission. Cyclophosphamide and rituximab have been showing encouraging results, whereas mycophenolate mofetil and azathioprine seem to be less effective [4]. Other therapeutic modalities that could be beneficial, especially in the presence of underlying CAPS, include plasma exchange and IVIg.

4.2 Adrenal hemorrhage

AH is a potentially devastating complication of aPL due to the resulting adrenal insufficiency. AH represents an infrequent cause of adrenal insufficiency, and besides aPL, it can be caused by other disorders, namely adrenal tumors and anatomical malformations, infections, and bleeding disorders (thrombocytopenia, heparin exposure) [59]. AH is a rare complication of aPL with its prevalence not precisely established. However, a significant proportion - one third - of affected patients have CAPS. The incidence in this subgroup is thus relatively high, between 10 to 16% [56]. A provoking moment usually initiates aPL-induced AH. Trauma, invasive procedures, infections, and warfarin withdrawal have been identified as such moments [60].

The main pathomechanism in aPL-induced AH, supported by the autopsy evidence, is multiple thromboses in the adrenal plexus leading to the secondary hemorrhage and destruction of the adrenal cortex. Due to its unique vascular anatomy (complex arterial system with three supplying arteries, rich vascular plexus in the zona reticularis, single drainage vein), the adrenal gland is prone to develop intraparenchymal hemorrhage in a case of venous obstruction. Vasculitis has not been found in aPL-induced AH [61].

AH usually manifests with back pain. Symptoms related to acute adrenal insufficiency (hypotension, malaise, fever, altered mental status, gastrointestinal symptoms including nausea, vomiting, and diarrhea) complement the clinical picture. Apart from the chronic adrenal insufficiency, skin hyperpigmentation is not present in the aPL-induced AH [59].

Laboratory tests and radiological imagining studies are critical for the confirmation of AH. Decreased cortisol levels and the lack of increase in cortisol levels after an adrenocorticotropic hormone stimulation test represent a typical laboratory finding. Abdominal contrast CT is the standard imagining method. However, CT has its limits, and if performed in the early phases of the bleeding, it may be falsely negative. A repeated CT scan is a must in the case of high clinical suspicion despite an initial negative result. Abdominal magnetic resonance is an alternative imagining method with the best imaging of the adrenal glands [62]. If the laboratory and imaging studies are inconclusive, adrenal biopsy remains the definitive diagnostic procedure. However, it is a high-risk procedure in terms of bleeding, and the risk–benefit ratio has to be evaluated individually. As a general rule, adrenal biopsy should be avoided in aPL-positive patients.

Clinical management has two goals: 1) to provide substitution of adrenal hormones, especially glucocorticoids; 2) to prevent further complications of aPL, namely thromboembolism. Since CAPS is a frequent finding in aPL-positive patients with AH, antithrombotics should be administered as long as possible despite hemorrhage. If their withdrawal is necessary, usually due to the extensive bleeding, the restart should be as soon as possible. The clinical experience stresses the critical importance of antithrombotic therapy. The study with aPL-positive patients and AH observed concurrent thrombotic events during the acute phase in 7 (43%) out of 16 participants. Five of 7 patients with confirmed thrombosis were diagnosed with CAPS [60]. Apart from the glucocorticoid substitution due to adrenal insufficiency, immunosuppressives are not a standard part of treatment since the available evidence does not confirm an effect on the clinical outcome [61]. However, their addition, alone or in combination with IVIg and plasma exchange, can be beneficial in the presence of CAPS.

The long-term prognosis of AH is relatively favorable after the acute phase, especially if antithrombotics are uninterrupted. In a review of 62 patients with AH followed for a mean of 25 (2–60) months, 90% (32 out of 35) of anticoagulated patients survived. Interestingly, overall mortality in the study reached 36% (25 out of 69 participants) [61]. Adrenal dysfunction is irreversible in most cases, although occasional recovery remains possible.

4.3 Catastrophic antiphospholipid syndrome

CAPS represents the most severe and potentially fatal variant of APS. It is characterized by excessive activation of hemostasis, rapid, multiple, and progressive thrombotic events, typically affecting small vessels, resulting in acute multiple organ dysfunction (usually kidneys, lungs, CNS, heart, skin) and TMAs [63]. Fortunately, CAPS is a relatively infrequent complication, affecting approximately 1% of patients with APS [2]. CAPS is the first manifestation of previously unrecognized or newly formed aPL in up to 50% of patients [10]. However, it can be the complication of preexisting and known aPL or APS as well. Its onset is usually - in about 2/3 of cases - related to precipitating factors such as infections, malignancies, trauma, invasive procedures, activation of underlying autoimmune disease, pregnancy complications, certain medications (oral contraceptives), and withdrawal or inadequate antithrombotic therapy. Pathological complement activation plays a critical role in its development [64].

Thromboembolic events and their complications dominate the clinical picture. Bleeding is typically secondary to the initial thromboembolism, although rarely can be among the initial clinical manifestations [65, 66]. The etiology of hemorrhage in CAPS is complex. It involves thrombocytopenia secondary to excessive platelet activation and consumption, consumption of coagulation factors, endothelial damage and dysfunction, thrombocytopenic thrombotic purpura (TTP)-like hemostatic changes, and development of DIC [67, 68]. Thrombocytopenia is a dominant change in CAPS, affecting up to 40% of patients with the complication. Thrombocytopenia, mainly if it manifests as the acute drop in platelet count in patients with aPL/APS and previously normal platelets, can be the first sign of impeding CAPS and precede the full clinical picture of CAPS for several days [11]. TTP-like changes frequently accompany thrombocytopenia [68]. Clinical presentation of hemorrhage is variable, with every organ system being a possible target. Life-threatening hemorrhage, including bleeding in the CNS and GIT, can occur [65, 66]. As mentioned before, DAH and AH are relatively frequent complications of CAPS.

The therapeutic approach is aggressive with several goals: 1) to suppress the immune system and production of aPL; 2) to prevent and treat thromboembolic events; 3) treat the underlying or provoking disorder. The combined immunosuppressive and immunomodulatory therapy (corticosteroids, IVIg, plasma exchange) together with full anticoagulation (preferably with heparin or LMWHs in the acute phase with the transition to warfarin) represents the initial therapeutic step [63]. Cyclophosphamide is the preferred immunosuppression in patients with underlying SLE. Rituximab and eculizumab are novel therapeutic possibilities that seem to be efficient in patients with predominant hematologic or microthromboangiopathic manifestations or resistant to first-line treatment [63, 69, 70]. Despite aggressive treatment and novel agents, the prognosis remains unfavorable in a significant number of cases, with a mortality rate reaching up to 40% [2]. The individual assessment of thrombotic and bleeding risk is an indispensable part of therapeutic management. The continuation of antithrombotic therapy is preferred over its tapering or withdrawal. Its continuation has to be considered even in the presence of hemorrhage.

5. Bleeding associated with antithrombotic agents

Bleeding events, particularly those involving the CNS and GIT, are regarded as potentially serious, but the expected adverse events of antithrombotic therapy. The incidence of major bleeding ranges from 3 to 6 per 100 person-years depending on the anticoagulant. It is high for patients on warfarin in particular [71]. The incidence of bleeding on antiplatelet therapy is generally lower, 3 to 4 per 1000 person-years [72]. The risk increases with the intensity of treatment or concomitant use of several agents. The combination of the anticoagulant with antiplatelet agent increases the risk of bleeding approximately 1.5 to 2-fold in comparison to anticoagulant therapy alone [73].

Since the presence of aPL represents a high-risk thrombophilia, antithrombotics – anticoagulants, antiplatelet agents, or their combination - are administered for a prolonged period, frequently life-long in most aPL-positive patients. The continuous administration of antithrombotic agents is used even in asymptomatic individuals with estimated high prothrombotic risk. Warfarin remains the preferred agent for anticoagulation, with the intent to achieve a higher INR range of 3.0–4.0 in specific clinical situations (recurrent thrombotic events, arterial events) [74].

Based on the current clinical practice and preferred intensity of therapy, aPL-positive patients receiving antithrombotics may seem to have an increased risk of treatment-related bleeding. However, available data show that hemorrhage does not represent the main clinical issue. The mortality rate due to thrombosis and its recurrence remains several times higher than the mortality rate related to bleeding. For example, a review of clinical studies documented 18 deaths related to recurrent thrombosis and only one due to hemorrhage [75]. The analysis of a prospective 10-year follow-up of 1000 patients with APS, performed as a part of the Euro-Phospholipid project, identified 34 deaths attributed to thromboembolism and only 10 to bleeding [76]. Reviews of clinical studies focused on anticoagulant therapy in APS suggest that, if INR on warfarin is within the standard therapeutic range, the major bleeding does not appear to be significantly more frequent in comparison to other patient groups on warfarin and is about 1.5–2.0% per year [77]. If higher INR levels (3.0–4.0) are needed, the risk of bleeding, but predominantly mild, increases significantly, approximately 2 to 2.5 times [77, 78]. As for antiplatelet agents, the rate of bleeding during their prophylactic or therapeutic use appears to be low, and major bleeding is rare [78, 79]. The risk of bleeding increases after invasive procedures, likely due to the use of bridging therapy, the early reintroduction of antithrombotics, and aggressive antithrombotic policies [80, 81]. Then again, thrombotic risk after surgery increases considerably as well despite preventive measures.

Independent predictors of major bleeding include overdose with warfarin (e.g. INR above 4.0), combined antithrombotic therapy, polypharmacy, age over 75 years, history of major bleeding (mostly gastrointestinal), malignancy, uncontrolled arterial hypertension, leukoaraiosis, and patient non-compliance [76–78]. It is critical to evaluate individual bleeding and prothrombotic risk and purposely identify potential risk factors. Caution is especially required when high-intensity anticoagulation or a combination of antithrombotics are indicated.

6. Conclusion

Bleeding is a rare but potentially severe complication of aPL and APS. Its etiology is heterogeneous; aPL-positive patients can develop bleeding due to thrombocytopenia, acquired coagulation factor deficiencies (predominantly hypoprothrombinemia), TMAs, or the adverse events of antithrombotic therapy (mostly with warfarin). However, thromboembolic events represent the most dangerous complications for aPL-positive patients, and the thrombotic risk remains clinically relevant even in the presence of hemorrhage in the majority of patients.

The management of bleeding is challenging. It is necessary to balance both thrombotic and bleeding stimuli and to continue antithrombotic prevention or therapy for as long as possible. The individual approach is critical for a favorable clinical outcome. Specific treatment can be necessary for eliminating the cause of bleeding and achieving its control. Immunosuppressive agents, especially corticosteroids, are the first-choice treatment for aPL-associated thrombocytopenia, coagulation factor deficiencies, CAPS, and DAH. Other immunosuppressive or immunomodulatory agents can be efficient in case of unsatisfactory clinical response. Rituximab appears to be the most promising alternative. Corticosteroids are also fundamental for the diffuse alveolar hemorrhage, albeit firstly for the correction of consequential adrenal insufficiency. aPL-positive patients receiving antithrombotics should be monitored closely, and their compliance ensured, especially in the scenario with the high-intensity or combined antithrombotic therapy.

Conflict of interest

The authors declare no conflict of interest.

Appendices and nomenclature

aCL	anticardiolipin antibodies
AH	adrenal hemorrhage
anti-B2GPI	anti-beta2-glycoprotein I antibodies
aPL	antiphospholipid antibodies
APS	antiphospholipid syndrome
aPTT	activated partial thromboplastin time
AWS	acquired von Willebrand syndrome
B2GPI	beta2-glycoprotein I
CAPS	catastrophic antiphospholipid syndrome
CNS	central nervous system
DAH	diffuse alveolar hemorrhage
DIC	disseminated intravascular coagulation
GIT	gastrointestinal tract
ITP	immune thrombocytopenia
IVIg	intravenous immunoglobulin
LA	lupus anticoagulant
LMWHs	low molecular weight heparins
PT	prothrombin time
SLE	systematic lupus erythematosus
TMAs	thrombotic microangiopathies
TPOMs	thrombopoietin mimetics
TTP	thrombocytopenic thrombotic purpura
vWF	von Willebrand factor

Author details

Peter Kubisz*, Pavol Holly and Jan Stasko
Department of Hematology and Transfusion Medicine, Jessenius Faculty of Medicine in Martin of the Comenius University in Bratislava, University Hospital in Martin, Martin, Slovakia

*Address all correspondence to: peter.kubisz@uniba.sk

IntechOpen

© 2021 The Author(s). Licensee IntechOpen. This chapter is distributed under the terms of the Creative Commons Attribution License (http://creativecommons.org/licenses/by/3.0), which permits unrestricted use, distribution, and reproduction in any medium, provided the original work is properly cited.

References

[1] Garcia D, Erkan D. The antiphospholipid syndrome. N Engl J Med. 2018;378:2010-2021. DOI: 10.1056/NEJMra1705454

[2] Abreu MM, Danowski A, Wahl DG, Amigo MC, Tektonidou M, Pacheco MS, Fleming N, Domingues V, Sciascia S, Lyra JO, Petri M, Khamashta M, Levy RA. The relevance of "non-criteria" clinical manifestations of antiphospholipid syndrome: 14th international congress on antiphospholipid antibodies. Autoimmun Rev. 2015;14:401-414. DOI: 10.1016/j.autrev.2015.01.002

[3] Xourgia E, Tektonidou MG. Management of non-criteria Manifestations in antiphospholipid syndrome. Current Rheumatology Reports. 2020;22:51. DOI: 10.1007/s11926-020-00935-2

[4] Pazzola G, Zuily S, Erkan D. The challenge of bleeding in antiphospholipid antibody-positive patients. Curr Rheumatol Rep. 2015;17:7. DOI: 10.1007/s11926-014-0481-0

[5] Cervera R, Boffa MC, Khamashta MA, Hughes GR. The euro-phospholipid project: Epidemiology of the antiphospholipid syndrome in Europe. Lupus. 2009;18:889-893. DOI: 10.1177/0961203309106832

[6] Uthman I, Godeau B, Taher A, Khamashta M. The hematologic manifestations of the antiphospholipid syndrome. Blood Rev. 2008;22:187-194. DOI: 10.1016/j.blre.2008.03.005

[7] Artim-Esen B, Diz-Küçükkaya R, İnanç M. The significance and management of thrombocytopenia in antiphospholipid syndrome. Curr Rheumatol Rep. 2015;17:14. DOI: 10.1007/s11926-014-0494-8

[8] Finazzi G, Brancaccio V, Moia M, Ciaverella N, Mazzucconi MG, Schinco PC, Ruggeri M, Pogliani EM, Gamba G, Rossi E, Baudo F, Manotti C, D'Angelo A, Palareti G, De Stefano V, Berrettini M, Barbui T. Natural history and risk factors for thrombosis in 360 patients with antiphospholipid antibodies: a four-year prospective study from the Italian registry. Am J Med. 1996;100:530-536. DOI: 10.1016/s0002-9343(96)00060-5

[9] Gómez-Puerta JA, Martin H, Amigo MC, Aguirre MA, Camps MT, Cuadrado MJ, Hughes GRV, Khamashta MA. Long-term follow-up in 128 patients with primary antiphospholipid syndrome: Do they develop lupus? Medicine (Baltimore). 2005;84:225-230. DOI: 10.1097/01.md.0000172074.53583.ea

[10] Rodríguez-Pintó I, Moitinho M, Santacreu I, Shoenfeld Y, Erkan D, Espinosa G, Cervera R, CAPS Registry Project Group (European Forum on Antiphospholipid Antibodies). Catastrophic antiphospholipid syndrome (CAPS): descriptive analysis of 500 patients from the international caps registry. Autoimmun Rev. 2016;15:1120-1124. DOI: 10.1016/j.autrev.2016.09.010

[11] Pontara E, Banzato A, Bison E, Cattini MG, Baroni G, Denas G, Calligaro A, Marson P, Tison T, Ruffatti A, Pengo V. Thrombocytopenia in high-risk patients with antiphospholipid syndrome. J Thromb Haemost. 2018;16:529-532. DOI:10.1111/jth.13947

[12] James TE, Martin LJ, Warkentin TE, Crowther MA. Catastrophic antiphospholipid syndrome refractory to high-dose intravenous immunoglobulin responsive to therapeutic plasma exchange. Platelets. 2020. DOI: 10.1080/09537104.2020.1802414

[13] Yelnik CM, Nguyen Y, Le Guern V, Hachulla E, Lambert M. Thrombocytopenia in primary antiphospholipid

syndrome, a marker of high-risk patients? Eur J Intern Med. 2020;74:106-107. DOI: 10.1016/j.ejim.2020.01.017.

[14] Abreu MM, Danowski A, Wahl DG, Amigo MC, Tektonidou M, Pacheco MS, Fleming N, Domingues V, Sciascia S, Lyra JO, Petri M, Khamashta M, Levy RA. The relevance of "non-criteria" clinical manifestations of antiphospholipid syndrome: 14th international congress on antiphospholipid antibodies. Autoimmun Rev. 2015;14:401-414. DOI: 10.1016/j.autrev.2015.01.002

[15] Demetrio Pablo R, Muñoz P, López-Hoyos M, Calvo V, Riancho L, Martínez-Taboada VM. Thrombocytopenia as a thrombotic risk factor in patients with antiphospholipid antibodies without disease criteria. Medicina Clínica (Barcelona). 2017;148:394-400. DOI: 10.1016/j.medcli.2016.11.026

[16] Comellas-Kirkerup L, Hernandez-Molina G, Cabral A. Antiphospholipid-associated thrombocytopenia or autoimmune hemolytic anemia in patients with or without definite primary antiphospholipid syndrome according to the Sapporo revised classification criteria: a 6-year follow-up study. Blood. 2010;116:3058-3063.

[17] Gansner JM, Berliner N. The rheumatology/hematology interface: CAPS and MAS diagnosis and management. Hematology Am Soc Hematol Educ Program. 2018;2018:313-317. DOI: 10.1182/asheducation-2018.1.313.

[18] Deucher A, Wool GD. How I investigate bone marrow necrosis. Int J Lab Hematol. 2019;41:585-592. DOI: 10.1111/ijlh.13091.

[19] Cruz ACL, Colella MP, De Paula EV, Annichinno-Bizzachi J, Orsi FA. Clinical course of primary immune thrombocytopenia with positive antiphospholipid antibodies. Eur J Intern Med 2019;69:e6–e7. DOI: 10.1016/j.ejim.2019.08.013

[20] Neunert C, Lim W, Crowther M, Cohen A, Solberg Jr L, Crowther MA. The American Society of Hematology 2011 evidence-based practice guideline for immune thrombocytopenia. Blood. 2011;117:4190-4207. DOI: 10.1182/blood-2010-08-302984

[21] Frison L, Lombardi A, Caputo I, Semenzato G, Fabris F, Vianello F. Relevance of antiphospholipid antibody profile in the clinical outcome of ITP: a single-Centre study. Hematology. 2019;24:134-138. DOI: 10.1080/10245332.2018.1532649

[22] Vreede AP, Bockenstedt PL, McCunea WJ, Knight JC. Cryptic conspirators: a conversation about thrombocytopenia and antiphospholipid syndrome. Curr Opin Rheumatol. 2019;31:231-240. DOI:10.1097/BOR.0000000000000595

[23] Neunert CE, Nichola Cooper N. Evidence-based management of immune thrombocytopenia: ASH guideline update. Hematology Am Soc Hematol Educ Program. 2018;2018:568-575. DOI: 10.1182/asheducation-2018.1.568

[24] Lidar M, Masarwa S, Rotman P, Carmi O, Rabinowicz N, Levy Y. Intravenous immunoglobulins for rheumatic disorders and thromboembolic events-a case series and review of the literature. Immunol Res. 2018;66:668-674. DOI: 10.1007/s12026-018-9047-y

[25] Erkan D, Vega J, Ramon G, Kozora E, Lockshin MD. A pilot open-label phase II trial of rituximab for non-criteria manifestations of antiphospholipid syndrome. Arthritis Rheum. 2013; 65:464-471. DOI: 10.1002/art.37759

[26] Sciascia S, Radin M, Cecchi I, Rubini E, Bazzan M, Roccatello D. Long-term effect of B-cells depletion alone as rescue therapy for severe thrombocytopenia in primary antiphospholipid syndrome. Semin Arthritis Rheum. 2019;48:741-744. DOI: 10.1016/j.semarthrit.2018.04.001

[27] Lusa A, Carlson A. Safety and efficacy of thrombopoeitin mimetics for refractory immune thrombocytopenia purpura in patients with systemic lupus erythematosus or antiphospholipid syndrome: a case Series. *lupus*. 2018; 27:1723-1728. DOI: 10.1177/0961203318770023

[28] Alkaabi JK, Alkindi S, Riyami NA, Zia F, Balla LMA, Balla SM. Successful treatment of severe thrombocytopenia with romiplostim in a pregnant patient with systemic lupus Erythematosus. *lupus*. 2012;21:1571-1574. DOI: 10.1177/0961203312463621

[29] LaMoreaux B, Barbar-Smiley F, Ardoin S, Madhoun H. Two cases of thrombosis in patients with antiphospholipid antibodies during treatment of immune thrombocytopenia with romiplostim, a thrombopoietin receptor agonist. Semin Arthritis Rheum. 2016;45:e10–e12. DOI: 10.1016/j.semarthrit.2015.07.008

[30] Gonzalez-Lopez TJ, Alvarez-Roman MT, Pascual C, Sánchez-González B, Fernández-Fuentes F, Pérez-Rus G, Hernández-Rivas JA, Bernat S, Bastida JM, Martínez-Badas MP, Martínez-Robles V, Soto I, Olivera P, Bolaños E, Alonso R, Entrena L, Gómez-Nuñez M, Alonso A, Cobo MY, Caparrós I, Tenorio M, Arrieta-Cerdán E, Lopez-Ansoar E, García-Frade J, González-Porras JR. Use of eltrombopag for secondary immune thrombocytopenia in clinical practice. Br J Haematol. 2017;178:959-970. DOI: 10.1111/bjh.14788

[31] Guitton Z, Terriou L, Lega JC, Nove-Josserand R, Hie M, Amoura Z, Bussel JB, Hamidou M, Rosenthal E, Lioger B, Chauveau D, Chaminade A, Magy-Bertrand N, Michel M, Audia S, Godeau B, Mahevas M. Risk of thrombosis with antiphospholipid syndrome in systemic lupus erythematosus treated with thrombopoietin receptor agonists. Rheumatology (Oxford). 2018; 57:1432-1438. DOI: 10.1093/rheumatology/key119

[32] Yu-Hin Chan E, Leung KFS, Soo ETL, Ma ALT. A Chinese Boy With Lupus Anticoagulant-hypoprothrombinemia Syndrome: A Case Report and Review of the Literature. J Pediatr Hematol Oncol. 2020 Apr 13. DOI: 10.1097/MPH.0000000000001805

[33] Mulliez SM, de Keyser F, Verbist C, Vantilborgh A, Wijns W, Beukinga I, Devreese KMJ. Lupus anticoagulant-hypoprothrombinemia syndrome: Report of two cases and review of the literature. Lupus. 2015;24:736-745. DOI: 10.1177/0961203314558859

[34] Sarah Vandamme Emilie Desclée Ver Elst K, Weekx S, Van Brusselen PMD, Vermeiren *S. lupus* anticoagulant hypoprothrombinaemia syndrome: An instructive paediatric case. J Paediatr Child Health. 2020 Apr 12. DOI: 10.1111/jpc.14889

[35] Forastiero R. Bleeding in the antiphospholipid syndrome. Hematology. 2012;17(suppl.1):s153-s155. DOI: 10.1179/102453312X13336169156654

[36] Sarker T, Roy S, Hollon W, Rajpurkar M. Lupus anticoagulant acquired hypoprothrombinemia syndrome in childhood: Two distinct patterns and review of the literature. Haemophilia. 2015;21:754-760. DOI: 10.1111/hae.12669

[37] Pilania RK, Suri D, Jindal AK, Kumar N, Sharma A, Sharma P, Guleria S, Rawat A, Ahluwalia J,

Singh S. Lupus anticoagulant hypoprothrombinemia syndrome associated with systemic lupus erythematosus in children: Report of two cases and systematic review of the literature. Rheumatol Int. 2018;38:1933-1940. DOI: 10.1007/s00296-018-4127-9

[38] Sreedharanunni S, Ahluwalia J, Kumar N, Bose SK, Dhawan R, Malhotra P. Lupus anticoagulant - hypoprothrombinemia syndrome: a rare cause of intracranial bleeding. Blood Coagul Fibrinolysis. 2017;28:416-418. DOI: 10.1097/MBC.0000000000000615

[39] Chen X, Nedved D, Plapp FV, Cunningham MT. Fatal pulmonary embolism and pulmonary hemorrhage in lupus anticoagulant hypoprothrombinemia syndrome: a case report and review of literature. Blood Coagul Fibrinolysis. 2018;29:708-713. DOI: 10.1097/MBC.0000000000000776

[40] Eddou H, Zinebi A, Khalloufi A, Mahtat EM, El Khadir SE, Moudden MK, Doghmi K, Mikdame M, El Baaj M. Thrombo-haemorrhagic disease-related hypoprothrombinemia-lupus anticoagulant syndrome revealing a light chains multiple myeloma. J Med Vasc. 2018;43:325-328. DOI: 10.1016/j.jdmv.2018.07.008

[41] Galland J, Mohamed S, Revuz S, de Maistre E, de Laat B, Marie PY, Zuily S, Lévy B, Regnault V, Wahl D. Lupus anticoagulant-hypoprothrombinemia syndrome and catastrophic antiphospholipid syndrome in a patient with antidomain I antibodies. Blood Coagul Fibrinolysis. 2016;27:580-582. DOI: 10.1097/MBC.0000000000000545

[42] Bajaj SP, Rapaport SI, Fierer DS, Herbst KD, Schwartz DB. A mechanism for hypoprothrombinemia of the acquired hypoprothrombinemia-lupus anticoagulant syndrome. Blood. 1983;61:684-692

[43] Ieko M, Yoshida M, Naito S, Ohmura K, Takahashi N. Lupus anticoagulant-hypoprothrombinemia syndrome and similar diseases: Experiences at a single center in Japan. Int J Hematol. 2019;110:197-204. DOI: 10.1007/s12185-019-02674-2

[44] Kobayashi N, Ogawa Y, Yanagisawa K, Ishizaki T, Uchiumi H, Suzuki N, Matsushita T, Ichinose A, Handa H. Acquired immune-mediated von Willebrand syndrome accompanied by antiphospholipid syndrome. Rinsho Ketsueki. 2017;58:613-618. DOI: 10.11406/rinketsu.58.613

[45] Gambichler T, Matip R. Erythromelalgia and livedo reticularis in a patient with essential Thrombocythemia, acquired von Willebrand disease, and elevated anti-phospholipid antibodies. Ann Dermatol. 2012;24:214-217. DOI: 10.5021/ad.2012.24.2.214

[46] Hanley D, Arkel YS, Lynch J, Kamiyama M. Acquired von Willebrand's syndrome in association with a lupus-like anticoagulant corrected by intravenous immunoglobulin. Am J Hematol. 1994;46:141-146. DOI: 10.1002/ajh.2830460216

[47] Dicke C, Holstein K, Schneppenheim S, Dittmer R, Schneppenheim R, Bokemeyer C, Iking-Konert C, Budde U, Langer F. Acquired hemophilia a and von Willebrand syndrome in a patient with late-onset systemic lupus erythematosus. Exp Hematol Oncol. 2014;3:21. DOI: 10.1186/2162-3619-3-21

[48] Viallard JF, Pellegrin JL, Vergnes C, Borel-Derlon A, Clofent-Sanchez G, Nurden AT, Leng B, Nurden P. Three cases of acquired von Willebrand disease associated with systemic lupus erythematosus. Br J Haematol. 1999;105:532-537. DOI: 10.1111/j.1365-2141.1999.01360

[49] Cao XY, Li MT, Zhang X, Zhao Y, Zeng XF, Zhang FC, Hou Y, Zhu LX.

Characteristics of acquired inhibitors to factor VIII and Von Willebrand factor secondary to systemic lupus erythematosus: Experiences from a Chinese tertiary medical center. J Clin Rheumatol. 2019 Dec 6. DOI: 10.1097/RHU.0000000000001284. Epub ahead of print.

[50] Gupta D, Chatterjee T, Sharma A, Ganguli P, Das S, Sharma S. Rare case of acquired Haemophilia and lupus anticoagulant. Indian J Hematol Blood Transfus. 2014;30:197-200. DOI: 10.1007/s12288-012-0204-5

[51] Bidot CJ, Jy W, Horstman LL, Huisheng H, Jimenez JJ, Yaniz M. Factor VII/VIIa: a new antigen in the anti-phospholipid antibody syndrome. Br J Haematol. 2003;120:618-626. DOI: 10.1046/j.1365-2141.2003.04161.x

[52] da Silva VA, Silva SS, Mendes Martins FF. Acquired deficiency of coagulation factor VII. Rev Bras Hematol Hemoter. 2015;37:269-271. DOI: 10.1016/j.bjhh.2015.05.002

[53] Lara AR, Schwarz MI. Diffuse alveolar hemorrhage. Chest. 2010;137:1164-1171. DOI: 10.1378/chest.08-2084

[54] Scheiman Elazary A, Cohen MJ, Aamar S, Dranitzki Z, Tayer-Shifman O, Mevorach D, Ben-Chetrit E, Naparstek Y. Pulmonary hemorrhage in antiphospholipid antibody syndrome. J Rheumatol. 2012;39:1628-1631. DOI: 10.3899/jrheum.120205

[55] Bucciarelli S, Espinosa G, Cervera R, Erkan D, Gómez-Puerta JA, Ramos-Casals M, Font J, Asherson RA, European forum on antiphospholipid antibodies. Mortality in the catastrophic antiphospholipid syndrome: Causes of death and prognostic factors in a series of 250 patients. Arthritis Rheum. 2006;54:2568-2576. DOI: 10.1002/art.22018

[56] Ogata Y, Fujieda Y, Sugawara M, Sato T, Ohnishi N, Kono M, Kato M, Oku K, Amengual O, Atsumi T. Morbidity and mortality in antiphospholipid syndrome based on cluster analysis: a 10-year longitudinal cohort study. Rheumatology (Oxford). 2020 Sep 18;keaa542. DOI: 10.1093/rheumatology/keaa542

[57] Stoots SA, Lief L, Erkan D. Clinical insights into diffuse alveolar hemorrhage in antiphospholipid syndrome. Curr Rheumatol Rep. 2019;21:56. DOI: 10.1007/s11926-019-0852-7

[58] Krause ML, Cartin-Ceba R, Specks U, Peikert T. Update on diffuse alveolar hemorrhage and pulmonary vasculitis. Immunol Allergy Clin North Am. 2012;32:587-600. DOI: 10.1016/j.iac.2012.08.001

[59] Charmandari E, Nicolaides NC, Chrousos GP. Adrenal insufficiency. Lancet. 2014;383:2152-2167. DOI: 10.1016/S0140-6736(13)61684-0

[60] Ramon I, Mathian A, Bachelot A, Hervier B, Haroche J, Boutin-Le Thi Huong D, Costedoat-Chalumeau N, Wechsler B, Karmali R, Velkeniers B, Touraine P, Coussieu C, Bennani A, Renard-Penna R, Grenier PA, Wahl D, Piette J-Ch, Amoura Z. Primary adrenal insufficiency due to bilateral adrenal hemorrhage-adrenal infarction in the antiphospholipid syndrome: Long-term outcome of 16 patients. J Clin Endocrinol Metab. 2013;98:3179-3189. DOI: 10.1210/jc.2012-4300

[61] Espinosa G, Santos E, Cervera R, Piette J-C, de la Red G, Gil V, Font J, Couch R, Ingelmo M, Asherson RA. Adrenal involvement in the antiphospholipid syndrome: Clinical and immunologic characteristics of 86 patients. Medicine (Baltimore). 2003;82:106-118.DOI:10.1097/00005792-200303000-00005

[62] Jordan E, Poder L, Courtier J, Sai V, Jung A, Coakley FV. Imaging of nontraumatic adrenal hemorrhage. AJR Am J Roentgenol. 2012;199:W91–W98. DOI: 10.2214/AJR.11.7973

[63] Cervera R, Rodríguez-Pintó I, Espinosa G. The diagnosis and clinical management of the catastrophic antiphospholipid syndrome: A comprehensive review. J Autoimmun. 2018;92:1-11. DOI: 10.1016/j.jaut.2018.05.007

[64] Chaturvedi S, Brodsky RA, McCrae KR. Complement in the pathophysiology of the antiphospholipid syndrome. Front Immunol. 2019;10:449. DOI: 10.3389/fimmu.2019.00449

[65] Adhikari A, Chisti MM, Bastola S, Kc O. Rare case of catastrophic antiphospholipid syndrome with spontaneous intracranial haemorrhage. BMJ Case Rep. 2019;12:e227171. DOI: 10.1136/bcr-2018-227171

[66] Jang SK, Sitek A, Said S, Shields RC. Catastrophic antiphospholipid syndrome as a cause of severe gastrointestinal bleeding. Am J Med. 2019;132:e661-e662. DOI: 10.1016/j.amjmed.2019.02.043

[67] Soe MH, Agarwal KA, Akough-Weir A. The wolf hidden behind the clots: Catastrophic antiphospholipid antibody syndrome. Case Rep Med. 2018;2018:4693037. DOI: 10.1155/2018/4693037

[68] Skoczynska M, Crowther MA, Chowaniec M, Ponikowska M, Chaturvedi S, Legault K. Thrombotic microangiopathy in the course of catastrophic antiphospholipid syndrome successfully treated with eculizumab: Case report and systematic review of the literature. Lupus. 2020;29:631-639. DOI: 10.1177/0961203320917460

[69] Tinti MG, Carnevale V, Inglese M, Molinaro F, Bernal M, Migliore A, De Cata A. Eculizumab in refractory catastrophic antiphospholipid syndrome: a case report and systematic review of the literature. Clin Exp Med. 2019;19:281-288. DOI: 10.1007/s10238-019-00565 8

[70] Unlu O, Erkan D. Catastrophic antiphospholipid syndrome: Candidate therapies for a potentially lethal disease. Annu Rev Med. 2017;68:287-296. DOI:10.1146/annurev-med-042915-102529

[71] Adeboyeje G, Sylwestrzak G, Barron JJ, White J, Rosenberg A, Abarca J, Crawford G, Redberg *R. major* Bleeding Risk During Anticoagulation with Warfarin, Dabigatran, Apixaban, or Rivaroxaban in Patients with Nonvalvular Atrial Fibrillation. J Manag Care Spec Pharm. 2017;23:968-978. DOI: 10.18553/jmcp.2017.23.9.968

[72] García Rodríguez LA, Martín-Pérez M, Hennekens CH, Rothwell PM, Lanas A. Bleeding risk with long-term low-dose aspirin: A systematic review of observational studies. PLoS One. 2016;11: e0160046. DOI: 10.1371/journal.pone.0160046

[73] Johnson SG, Rogers K, Delate T, Witt DM. Outcomes associated with combined antiplatelet and anticoagulant therapy. Chest. 2008;133:948-954. DOI: 10.1378/chest.07-2627

[74] Ghembaza A, Saadoun D. Management of Antiphospholipid Syndrome. Biomedicines. 2020;8:508. DOI: 10.3390/biomedicines8110508

[75] Ruiz-Irastorza G, Hunt BJ, Khamashta MA. A systematic review of secondary thromboprophylaxis in patients with antiphospholipid antibodies. Arthritis Rheum. 2007;57:1487-1495. DOI: 10.1002/art.23109

[76] Cervera R, Serrano R, Pons-Estel GJ, Ceberio-Hualde L, Shoenfeld Y, de Ramón E, Buonaiuto V, Jacobsen S, Zeher MM, Tarr T, Tincani A, Taglietti M, Theodossiades G, Nomikou E, Galeazzi M, Bellisai F, Meroni PL, Derksen RHWD, de Groot PGD, Baleva M, Mosca M, Bombardieri S, Houssiau F, Gris JC, Quéré I, Hachulla E, Vasconcelos C, Fernández-Nebro A, Haro M, Amoura Z, Miyara M, Tektonidou M, Espinosa G, Bertolaccini ML, Khamashta MA, Euro-Phospholipid Project Group (European Forum on Antiphospholipid Antibodies). Morbidity and mortality in the antiphospholipid syndrome during a 10-year period: a multicentre prospective study of 1000 patients. Ann Rheum Dis. 2015;74:1011-1018. DOI: 10.1136/annrheumdis-2013-204838

[77] Ruiz-Irastorza G, Khamashta MA, Hunt BJ, Escudero A, Cuadrado MJ, Hughes GR. Bleeding and recurrent thrombosis in definite antiphospholipid syndrome: analysis of a series of 66 patients treated with oral anticoagulation to a target international normalized ratio of 3.5. Arch Intern Med. 2002;162:1164-1169. DOI: 10.1001/archinte.162.10.1164

[78] Bala MM, Celinska-Lowenhoff M, Szot W, Padjas A, Kaczmarczyk M, Swierz MJ, Undas A. Antiplatelet and anticoagulant agents for secondary prevention of stroke and other thromboembolic events in people with antiphospholipid syndrome. Cochrane Database Syst Rev. 2020 Oct 12;10:CD012169. DOI: 10.1002/14651858.CD012169.pub3

[79] MM, Paszek E, Lesniak W, Wloch-Kopec D, Jasinska K, Undas A. Antiplatelet and anticoagulant agents for primary prevention of thrombosis in individuals with antiphospholipid antibodies. Cochrane Database Syst Rev. 2018;7:CD012534. DOI: 10.1002/14651858.CD012534.pub2

[80] Atisha-Fregoso Y, Espejo-Poox E, Carrillo-Maravilla E, Pulido-Ramírez AL, Lugo Baruqui D, Hernández-Molina G, Cabral AR. Perioperative management of patients with antiphospholipid syndrome: a single-center experience. Rheumatol Int. 2017;37:1159-1164. DOI: 10.1007/s00296-017-3727-0

[81] Chen RX, Zhou YZ, Li PC, Yang HX, Fei YY, Hu XM, Wu W, Zhao LD, Chen H, Zhang X. The efficacy and safety of antithrombotic therapy in patients with positive antiphospholipid antibodies receiving invasive procedures: Experience from a single tertiary center. Clin Rheumatol. 2019;38:1897-1904. DOI: 10.1007/s10067-019-04483-8

Section 3

Trends and Novelties in the Diagnosis and Pathogenesis of Antiphospholipid Syndrome

Chapter 5

A Novel Autoantibody against β2-Glycoprotein I/HLA Class II Complexes in Antiphospholipid Syndrome

Kenji Tanimura, Yuki Sasagawa, Masashi Deguchi, Noriko Arase, Hisashi Arase and Hideto Yamada

Abstract

We have found that a novel autoantibody against β2-glycoprotein I (β2GPI)/human leukocyte antigen (HLA) class II complexes (anti-β2GPI/HLA-DR) is involved in the pathogenesis of antiphospholipid syndrome (APS). It was also found that many APS patients who were negative for conventional antiphospholipid antibodies (aPLs) possessed anti-β2GPI/HLA-DR. These results suggested that anti-β2GPI/HLA-DR measurements may be more sensitive for diagnosing APS than conventional aPLs tests. Recurrent pregnancy loss (RPL) is one of the clinical manifestations of APS. Therefore, a prospective, multicenter, cross-sectional study were conducted to assess whether anti-β2GPI/HLA-DR is also associated with RPL. This study of 227 couples with RPL revealed that 22.9% (52/227) of RPL women tested positive for anti-β2GPI/HLA-DR, and 24 (19.8%) of the 121 couples with unexplained RPL tested positive for anti-β2GPI/HLA-DR. Interestingly, thirty-five of the 52 (67.3%) RPL patients who were positive for anti-β2GPI/HLA-DR possessed no conventional aPLs of criteria. This novel autoantibody against β2GPI/HLA class II complexes may be a major risk factor for RPL, and it may be a promising biomarker for diagnosing APS.

Keywords: Autoantibody, β2-glycoprotein I, HLA class II, recurrent pregnancy loss

1. Introduction

It is well known that specific human leukocyte antigen (HLA) class II alleles are associated with susceptibility to many autoimmune diseases [1]. However, the mechanisms by which specific HLA class II molecules control the immune response in autoimmune diseases have been unclear. On the other hand, autoantibodies are produced in most autoimmune diseases and cause clinical manifestations of the diseases. It has also been an enigma how autoantibodies targeting self-antigens cause the autoimmune diseases. Arase *et al.* discovered a novel function of HLA class II molecules which are involved in the pathogenesis of certain autoimmune diseases [2–5].

This review will focus on the autoantibodies associating with the novel function of HLA class II molecules and the pathogenesis of antiphospholipid syndrome (APS).

2. The novel function of HLA class II molecules and autoimmune diseases

The classical function of HLA class II molecules is to present antigen peptides, derived from exogeneous proteins digested in lysosomes, to helper T-cells and by that to activate them.

Endogenous proteins, on the other hand, are formed and folded in the endoplasmic reticulum (ER). Correctly folded proteins are essential for cell survival and function. Therefore, it is believed that misfolded proteins generated in the ER are never transported to the extracellular space, because such proteins are eliminated by ER-associated degradation (ERAD).

However, Arase *et al.* discovered that misfolded proteins can be rescued from ERAD and transported to the cell surface without being processed into peptides. This process occurs in the ER via an association between the misfolded proteins and the peptide-binding groove of HLA class II molecules [2].

In addition, misfolded proteins complexed with HLA class II molecules of disease-susceptible alleles have been found to serve as targets of autoantibodies in certain autoimmune diseases, and to be involved in the disease pathogenesis. For example, immunoglobulin (Ig) G heavy chain complexed with HLA-DR and myeloperoxidase complexed with HLA-DR are major targets for autoantibodies in patients with rheumatoid arthritis and microscopic polyangiitis, respectively [3, 5].

3. The conventional concepts of antiphospholipid antibodies in APS

APS is diagnosed both by the presence of clinical manifestations, including vascular thrombosis and pregnancy morbidity, and by the presence of antiphospholipid antibodies (aPLs) which present a laboratory criteria for APS [6]. Laboratory criteria for APS include IgG and IgM anticardiolipin antibodies (aCLs), IgG and IgM anti-β2-glycoprotein I (aβ2GPI) antibodies, and lupus anticoagulant (LAC). aPLs are thought to recognize linear β2-glycoprotein I (β2GPI), which undergoes conformational changes from the circular form of β2GPI by binding to negatively charged phospholipids [7], and cause APS by interacting with vascular endothelial cells [8]. Therefore, β2GPI bound to negatively charged phospholipids or negatively charged plates is used clinically to detect autoantibodies in APS patients [9]. However, because autoantibodies against the β2GPI complexed to negatively charged phospholipids or high binding plates are detected in less than half of patients with clinical manifestations of APS [10–12], these facts suggest that additional targets of autoantibodies may exist. Furthermore, because β2GPI is a secreted protein, it cannot be universally present on the cell surface. Therefore, there might be other specific molecules which present β2GPI on the surface of vascular endothelial cells.

4. The discovery of a novel autoantibody against β2GPI/HLA-DR complex in APS

We found that 293 T cells co-transfected with β2GPI and HLA-DR expressed both β2GPI and HLA-DR on the cell surface by flow cytometry analysis

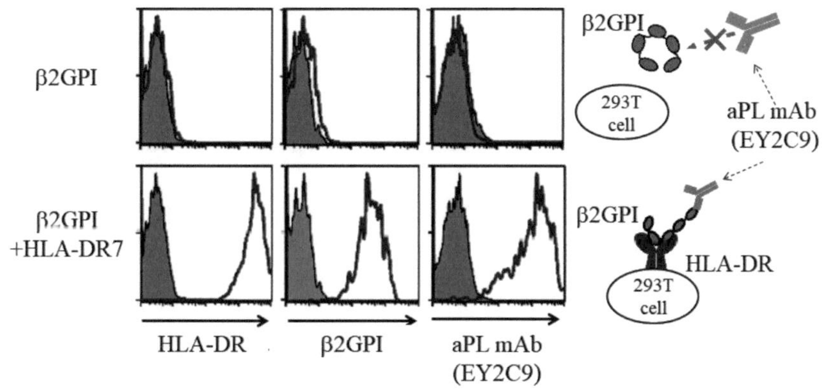

Figure 1.
Monoclonal anti-phospholipid antibody binds to β2GPI/HLA-DR complex on the cell surface. 293 T cells transfected with only β2GPI did not express β2GPI on the cell surface, and human monoclonal anti-phospholipid antibody (EY2C9) did not bind to these cells (the upper 3 histograms and 1 scheme). When β2GPI was co-transfected with HLA-DR into 293 T cells, β2GPI was expressed on the cell surface and was recognized by EY2C9 monoclonal antibody (the lower 3 histograms and 1 scheme). Abbreviations: HLA, human leukocyte antigen; β2GPI, β2-glycoprotein I; aPL mAb, anti-phospholipid monoclonal antibody.

(**Figure 1**) [4]. Conversely, 293 T cells transfected with only β2GPI did not express β2GPI on the cell surface, because β2GPI is a secreted protein (**Figure 1**) [4]. Immunoprecipitation and immunoblotting experiments revealed that full-length β2GPI proteins, but not peptide fragments of β2GPI, formed a complex with HLA-DR, and that these full-length β2GPI/HLA-DR complexes were present on the cell surface [4].

Furthermore, flow cytometry analysis revealed that not only the monoclonal antiphospholipid antibody derived from an APS patient (EY2C9), but also antibodies in the sera of APS patients can bind to the β2GPI/HLA-DR complexes, even in the absence of phospholipids [4].

5. Autoantibodies targeting β2GPI/HLA-DR complex are involved in the pathogenesis of APS

Immunofluorescence staining and *in situ* proximity-ligation assay (PLA), which detect close proximity (less than 40 nm) between two molecules [13], showed that β2GPI and HLA-DR were co-localized in endothelial cells of the placental decidua vessels from APS patients with spontaneous abortion. In contrast, no co-localization of β2GPI and HLA-DR was observed in placental tissues obtained from patients without APS [4].

In addition, we found that monoclonal antibody EY2C9 exhibited complement-mediated cytotoxicity against 293 T cells expressing β2GPI together with the APS susceptibility allele HLA-DR7, however the cytotoxicity was not detected against 293 T cells expressing HLA-DR7 alone or against those transfected with β2GPI alone [4].

HLA class II expression on endothelial cells is known to be induced after exposure to cytokines, such as IFN-γ and TNF-α [14]. Therefore, inflammatory stimuli can induce HLA class II expression on vascular endothelial cells, and HLA class II molecules transport structurally altered β2GPI, which has high affinity for the peptide-binding grooves of the alleles of HLA class II. Autoantibodies against β2GPI/HLA class II complexes may damage vascular endothelial cells expressing β2GPI/HLA class II complexes in a complement-dependent manner and cause clinical manifestations of APS, including vascular thrombosis and pregnancy complications. In this way, β2GPI/HLA class II complexes and autoantibodies against the complexes may be involved in the pathogenesis of APS.

6. Alleles of HLA-DR complexed with β2GPI affect susceptibility to APS

HLA-DR4, HLA-DR7, and HLA-DR13 have been reported as susceptibility alleles for APS [15–18]. However, the mechanism by which these HLA class II alleles increase susceptibility to APS has remained an enigma.

To address this issue, we analyzed the ability of different HLA-DR alleles to transport β2GPI to the cell surface and found that HLA-DR7 and HLA-DR4 could transport much higher levels of β2GPI than other HLA-DR alleles recognized by the EY2C9 monoclonal antibody [4]. These results indicated that a binding affinity of β2GPI to each HLA-DR allele is important for autoantibody recognition of β2GPI/HLA-DR complexes and is associated with differences in susceptibility to APS between different HLA-DR alleles.

7. A method for quantifying serum levels of autoantibodies against β2GPI/HLA-DR complexes

We developed and modified a method to measure serum levels of autoantibodies against β2GPI/HLA-DR complexes (anti-β2GPI/HLA-DR) [4, 19].

Green fluorescent protein (GFP)-labeled β2GPI/HLA-DR complex-expressing 293 T cells and DsRed-labeled HLA-DR-expressing 293 T cells were generated by transient transfection [19]. A serum sample from a patient in whom anti-β2GPI/HLA-DR were detectable after a 10^6-fold dilution was used as a standard serum. The anti-β2GPI/HLA-DR level of a standard serum was defined as 1,000 units. The mean fluorescence intensity (MFI) of IgG binding to transfected cells in the sample sera was analyzed by flow cytometry. Specific IgG binding to the β2GPI/HLA-DR complex was calculated by subtracting the MFI of IgG binding to HLA-DR-expressing cells from β2GPI/HLA-DR complex-expressing cells. Serum levels of anti-β2GPI/HLA-DR in each sample were calculated from the standard curve generated by measuring specific IgG binding to the β2GPI/HLA-DR complex in serially diluted standard serum.

8. Autoantibody against β2GPI/HLA-DR complex is a promising novel biomarker for APS

In our previous study, we measured serum levels of anti-β2GPI/HLA-DR in stored sera from 120 patients with APS, most of whom had a history of vascular thrombosis, and found that 83% of the 120 patients had autoantibodies directed against β2GPI/HLA-DR complexes. Furthermore, about 50% of the APS patients who tested positive for anti-β2GPI/HLA-DR (< 99th percentile values measured in sera of 100 healthy subjects) were negative for both IgG aCLs and IgG aβ2GPI antibodies [4]. Another recent study also showed that 27% of 111 patients with idiopathic chronic limb ulcers who were negative for aPLs possessed anti-β2GPI/HLA-DR [20]. These results suggest that anti-β2GPI/HLA-DR are associated with APS manifestations, even in patients who do not meet the diagnostic criteria for APS because they are negative for conventional aPLs.

The latest prospective, multicenter, cross-sectional study, of 227 couples with recurrent pregnancy loss (RPL), which is one of the clinical manifestations of APS, revealed that 22.9% (52/227) of women with RPL tested positive for anti-β2GPI/HLA-DR (< 99th percentile values measured in sera of 208 healthy, fertile control women) [19]. In this study, anti-β2GPI/HLA-DR were detected most frequently in women with RPL among other commonly recognized risk factors for RPL, i.e., uterine malformation, thyroid dysfunction, chromosomal abnormality, aPLs positive, low factor XII activity, low protein S activity, and low protein C activity (**Figure 2**). Importantly, 53.3% (121/227) of women with RPL had no commonly accepted risk factors for RPL, and 24 of these 121 (19.8%) women with unexplained RPL were positive for anti-β2GPI/HLA-DR (**Figure 2**). In addition, 45 of the 227 women with RPL (19.8%) were positive for at least one of the 5 conventional aPLs meeting the diagnostic criteria for APS in this study, i.e., IgG aCL (8.8%), IgM aCL (6.2%), IgG aβ2GPI (3.1%), IgM aβ2GPI (1.3%), and LAC (2.6%). The rate of positivity for anti-β2GPI/HLA-DR was the highest (22.9%) of the 5 aPLs that met the diagnostic criteria for APS. Notably, 35 (67.3%) of the 52 women with RPL who were positive for anti-β2GPI/HLA-DR, were negative for APS laboratory criteria (**Figure 3**).

On the other hand, the presence of multiple aPLs and LAC positivity has been reported to be strongly associated with the severity of clinical manifestations of APS [21–26]. In our study, all 3 women with RPL who had double or triple aPLs positivity were also positive for anti-β2GPI/HLA-DR, and the 2 with triple positivity had very high anti-β2GPI/HLA-DR levels (927.5 units and 330.7 units). First of both women experienced early-onset HELLP syndrome (hemolysis, elevated liver enzymes, and low platelets) at 14 weeks of gestation, and the second experienced a thromboembolism with cerebral infarction [19]. Multiple positivity for aPLs may be associated with higher levels of anti-β2GPI/HLA-DR, and these conditions may be closely associated with the severity of the clinical manifestations of APS.

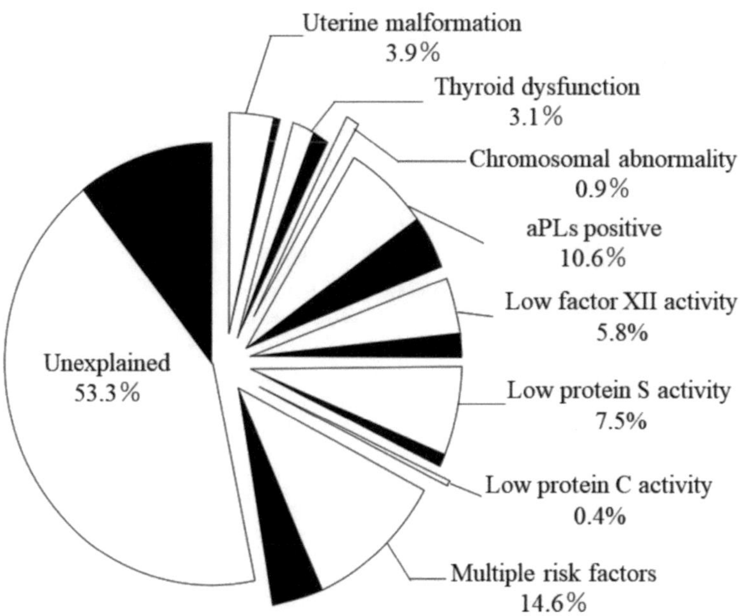

Figure 2.
Risk factors for recurrent pregnancy loss (RPL) among 227 women with RPL. All women with RPL enrolled in this study attended evaluations to identify commonly accepted risk factors for RPL. Black pie slices indicate the frequencies of women with RPL who were also positive for anti-β2GPI/HLA-DR (n = 52). Abbreviations: aPLs, antiphospholipid antibodies.

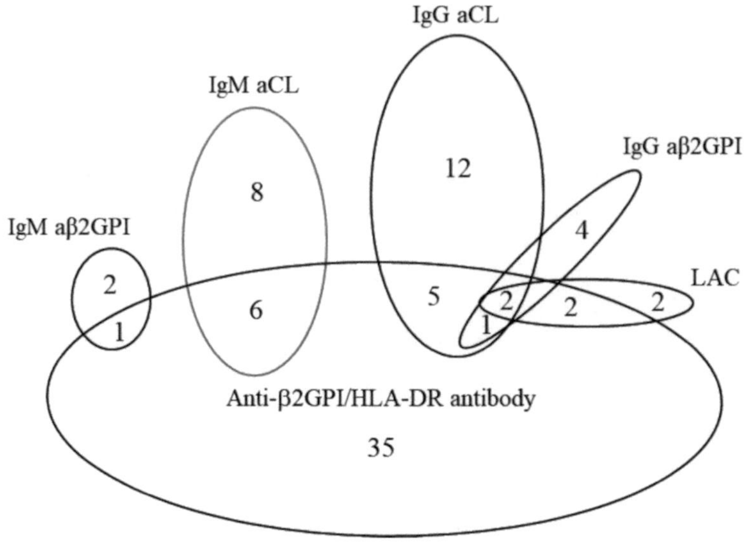

Figure 3.
Positivity for anti-β2-glycoprotein I /HLA-DR antibodies (anti-β2GPI/HLA-DR) and antiphospholipid antibodies (aPLs) in 227 women with recurrent pregnancy loss (RPL). Numbers in the Venn diagram represent the number of women who had unique or nonunique results in tests for aPLs and anti-β2GPI/HLA-DR. abbreviations: Ig, immunoglobulin; HLA, human leukocyte antigen; β2GPI, β2-glycoprotein I; aβ2GPI, anti-β2-glycoprotein I antibody; aCL, anti-cardiolipin antibody; LAC, lupus anticoagulant.

9. The future perspectives of the clinical use of autoantibodies targeting β2GPI/HLA-DR complexes

The standard treatment for pregnant women with APS is combination therapy with heparin and low-dose aspirin (LDA) [27], and the same therapy could also be effective in the treatment of women with RPL and anti-β2GPI/HLA-DR positivity. A cohort study is already underway to assess the efficacy of LDA and/or heparin therapy in such women. The history of vascular thrombosis and obstetric complications, including hypertensive disorders of pregnancy and fetal growth restriction, has not been investigated in prospective studies. Future studies assessing whether anti-β2GPI/HLA-DR are associated with thrombosis, hypertensive disorders of pregnancy, and fetal growth restriction are needed.

Further understanding of these novel autoantibodies associated with novel function of HLA class II molecules will provide new insights into the etiology of not only APS but also other autoimmune diseases and might lead to development of new treatment strategies for these diseases.

Author details

Kenji Tanimura[1,2], Yuki Sasagawa[1,2], Masashi Deguchi[1], Noriko Arase[2], Hisashi Arase[2] and Hideto Yamada[1*]

1 Department of Obstetrics and Gynecology, Kobe University Graduate School of Medicine, Kobe, Japan

2 Department of Immunochemistry, Research Institute for Microbial Disease, Osaka University, Suita, Japan

*Address all correspondence to: yhideto@med.kobe-u.ac.jp

IntechOpen

© 2021 The Author(s). Licensee IntechOpen. This chapter is distributed under the terms of the Creative Commons Attribution License (http://creativecommons.org/licenses/by/3.0), which permits unrestricted use, distribution, and reproduction in any medium, provided the original work is properly cited.

References

[1] Jones EY, Fugger L, Strominger JL, Siebold C. 2006. MHC class II proteins and disease: a structural perspective. *Nat Rev Immunol* **6:**271-282.

[2] Jiang Y, Arase N, Kohyama M, Hirayasu K, Suenaga T, Jin H, Matsumoto M, Shida K, Lanier LL, Saito T, Arase H. 2013. Transport of misfolded endoplasmic reticulum proteins to the cell surface by MHC class II molecules. *Int Immunol* **25:**235-246.

[3] Jin H, Arase N, Hirayasu K, Kohyama M, Suenaga T, Saito F, Tanimura K, Matsuoka S, Ebina K, Shi K, Toyama-Sorimachi N, Yasuda S, Horita T, Hiwa R, Takasugi K, Ohmura K, Yoshikawa H, Saito T, Atsumi T, Sasazuki T, Katayama I, Lanier LL, Arase H. 2014. Autoantibodies to IgG/HLA class II complexes are associated with rheumatoid arthritis susceptibility. *Proc Natl Acad Sci U.S.A* **111:**3787-3792.

[4] Tanimura K, Jin H, Suenaga T, Morikami S, Arase N, Kishida K, Hirayasu K, Kohyama M, Ebina Y, Yasuda S, Horita T, Takasugi K, Ohmura K, Yamamoto K, Katayama I, Sasazuki T, Lanier LL, Atsumi T, Yamada H, Arase H. 2015. β2-Glycoprotein I/HLA class II complexes are novel autoantigens in antiphospholipid syndrome. *Blood* **125:**2835-2844.

[5] Hiwa R, Ohmura K, Arase N, Jin H, Hirayasu K, Kohyama M, Suenaga T, Saito F, Terao C, Atsumi T, Iwatani H, Mimori T, Arase H. 2017. Myeloperoxidase/HLA Class II Complexes Recognized by Autoantibodies in Microscopic Polyangiitis. *Arthritis Rheumatol* **69:**2069-2080.

[6] Miyakis S, Lockshin MD, Atsumi T, Branch DW, Brey RL, Cervera R, Derksen RH, PG DEG, Koike T, Meroni PL, Reber G, Shoenfeld Y, Tincani A, Vlachoyiannopoulos PG, Krilis SA. 2006. International consensus statement on an update of the classification criteria for definite antiphospholipid syndrome (APS). *J Thromb Haemost* **4:**295-306.

[7] Agar C, van Os GM, Morgelin M, Sprenger RR, Marquart JA, Urbanus RT, Derksen RH, Meijers JC, de Groot PG. 2010. β2-glycoprotein I can exist in 2 conformations: implications for our understanding of the antiphospholipid syndrome. *Blood* **116:**1336-1343.

[8] Giannakopoulos B, Krilis SA. 2013. The pathogenesis of the antiphospholipid syndrome. *N Engl J Med* **368:**1033-1044.

[9] Amengual O, Atsumi T, Khamashta MA, Koike T, Hughes GR. 1996. Specificity of ELISA for antibody to β2-glycoprotein I in patients with antiphospholipid syndrome. *Br J Rheumatol* **35:**1239-1243.

[10] Atsumi T, Ieko M, Bertolaccini ML, Ichikawa K, Tsutsumi A, Matsuura E, Koike T. 2000. Association of autoantibodies against the phosphatidylserine-prothrombin complex with manifestations of the antiphospholipid syndrome and with the presence of lupus anticoagulant. *Arthritis Rheum* **43:**1982-1993.

[11] Otomo K, Atsumi T, Amengual O, Fujieda Y, Kato M, Oku K, Horita T, Yasuda S, Koike T. 2012. Efficacy of the antiphospholipid score for the diagnosis of antiphospholipid syndrome and its predictive value for thrombotic events. *Arthritis Rheum* **64:**504-512.

[12] Gardiner C, Hills J, Machin SJ, Cohen H. 2013. Diagnosis of antiphospholipid syndrome in routine clinical practice. *Lupus* **22:**18-25.

[13] Soderberg O, Gullberg M, Jarvius M, Ridderstrale K, Leuchowius KJ, Jarvius J, Wester K, Hydbring P, Bahram F, Larsson LG, Landegren U. 2006. Direct observation of individual endogenous protein complexes in situ by proximity ligation. *Nat Methods* **3:**995-1000.

[14] Collins T, Korman AJ, Wake CT, Boss JM, Kappes DJ, Fiers W, Ault KA, Gimbrone MA, Jr., Strominger JL, Pober JS. 1984. Immune interferon activates multiple class II major histocompatibility complex genes and the associated invariant chain gene in human endothelial cells and dermal fibroblasts. *Proc Natl Acad Sci U.S.A* **81:**4917-4921.

[15] Savi M, Ferraccioli GF, Neri TM, Zanelli P, Dall'Aglio PP, Tincani A, Balestrieri G, Carella G, Cattaneo R. 1988. HLA-DR antigens and anticardiolipin antibodies in northern Italian systemic lupus erythematosus patients. *Arthritis Rheum* **31:**1568-1570.

[16] Hartung K, Coldewey R, Corvetta A, Deicher H, Kalden JR, Krapf F, Lang B, Lakomek HJ, Liedvogel B, Peter HH, et al. 1992. MHC gene products and anticardiolipin antibodies in systemic lupus erythematosus results of a multicenter study. SLE Study Group. *Autoimmunity* **13:**95-99.

[17] Domenico Sebastiani G, Minisola G, Galeazzi M. 2003. HLA class II alleles and genetic predisposition to the antiphospholipid syndrome. *Autoimmun Rev* **2:**387-394.

[18] Granados J, Vargas-Alarcon G, Drenkard C, Andrade F, Melin-Aldana H, Alcocer-Varela J, Alarcon-Segovia D. 1997. Relationship of anticardiolipin antibodies and antiphospholipid syndrome to HLA-DR7 in Mexican patients with systemic lupus erythematosus (SLE). *Lupus* **6:**57-62.

[19] Tanimura K, Saito S, Nakatsuka M, Nagamatsu T, Fujii T, Fukui A, Deguchi M, Sasagawa Y, Arase N, Arase H, Yamada H. 2020. The β2 -Glycoprotein I/HLA-DR Complex As A Major Autoantibody Target in Obstetric Antiphospholipid Syndrome. *Arthritis Rheumatol* **72:** 1882-1891.

[20] Arase N, Tanimura K, Jin H, Yamaoka T, Kishibe M, Nishioka M, Kiyohara E, Tani M, Matsuoka S, Ohmura K, Takasugi K, Yamamoto T, Murota H, Arase H, Katayama I. 2018. Novel autoantibody against the β2-glycoprotein I/human leucocyte antigen-DR complex in patients with refractory cutaneous ulcers. *Br J Dermatol* **178:**272-275.

[21] Pengo V, Ruffatti A, Legnani C, Testa S, Fierro T, Marongiu F, De Micheli V, Gresele P, Tonello M, Ghirarduzzi A, Bison E, Denas G, Banzato A, Padayattil Jose S, Iliceto S. 2011. Incidence of a first thromboembolic event in asymptomatic carriers of high-risk antiphospholipid antibody profile: a multicenter prospective study. *Blood* **118:**4714-4718.

[22] Ruffatti A, Del Ross T, Ciprian M, Bertero MT, Sciascia S, Scarpato S, Montecucco C, Rossi S, Caramaschi P, Biasi D, Doria A, Rampudda M, Monica N, Fischetti F, Picillo U, Brucato A, Salvan E, Vittorio P, Meroni P, Tincani A, Antiphospholipid Syndrome Study Group of Italian Society of R. 2011. Risk factors for a first thrombotic event in antiphospholipid antibody carriers: a prospective multicentre follow-up study. *Ann Rheum Dis* **70:**1083-1086.

[23] Yamada H, Atsumi T, Kato EH, Shimada S, Morikawa M, Minakami H. 2003. Prevalence of diverse antiphospholipid antibodies in women with recurrent spontaneous abortion. *Fertil Steril* **80:**1276-1278.

[24] Galli M, Luciani D, Bertolini G, Barbui T. 2003. Lupus anticoagulants are

stronger risk factors for thrombosis than anticardiolipin antibodies in the antiphospholipid syndrome: a systematic review of the literature. *Blood* **101:**1827-1832.

[25] Lockshin MD, Kim M, Laskin CA, Guerra M, Branch DW, Merrill J, Petri M, Porter TF, Sammaritano L, Stephenson MD, Buyon J, Salmon JE. 2012. Prediction of adverse pregnancy outcome by the presence of lupus anticoagulant, but not anticardiolipin antibody, in patients with antiphospholipid antibodies. *Arthritis Rheum* **64:**2311-2318.

[26] Gebhart J, Posch F, Koder S, Perkmann T, Quehenberger P, Zoghlami C, Ay C, Pabinger I. 2015. Increased mortality in patients with the lupus anticoagulant: the Vienna Lupus Anticoagulant and Thrombosis Study (LATS). *Blood* **125:**3477-3483.

[27] RPL EGGo, Bender Atik R, Christiansen OB, Elson J, Kolte AM, Lewis S, Middeldorp S, Nelen W, Peramo B, Quenby S, Vermeulen N, Goddijn M. 2018. ESHRE guideline: recurrent pregnancy loss. *Hum Reprod Open* **2018:**hoy004.

Chapter 6

Extracellular Vesicles: Intercellular Communication Mediators in Antiphospholipid Syndrome

Ula Štok, Saša Čučnik, Snežna Sodin-Šemrl and Polona Žigon

Abstract

Antiphospholipid syndrome (APS) is a systemic autoimmune disease characterized by thrombosis, obstetric complications and the presence of antiphospholipid antibodies (aPL) that cause endothelial injury and thrombophilia. Extracellular vesicles are involved in endothelial and thrombotic pathologies and may therefore have an influence on the prothrombotic status of APS patients. Intercellular communication and connectivity are important mechanisms of interaction between healthy and pathologically altered cells. Despite well-characterized *in vitro* and *in vivo* models of APS pathology, the field of extracellular vesicles is still largely unexplored and could therefore provide an insight into the APS mechanism and possibly serve as a biomarker to identify patients at increased risk. The analysis of EVs poses a challenge due to the lack of standardized technology for their isolation and characterization. Recent findings in the field of EVs offer promising aspects that may explain their role in the pathogenesis of various diseases, including APS.

Keywords: Extracellular vesicles, Antiphospholipid syndrome, Antiphospholipid antibodies, Thrombosis, Extracellular vesicles, Endothelial cells, Monocytes, Platelets

1. Introduction

Antiphospholipid syndrome (APS) is a systemic autoimmune disease characterized by thrombosis and/or obstetric complications and persistent presence of antiphospholipid antibodies (aPL) [1]. aPL cause the activation of cells involved in the vasculature (endothelial cells, platelets, monocytes) and the release of extracellular vesicles (EVs). EVs are submicron particles that are constitutively released from nearly all cell types [2] and circulate in plasma of healthy individuals in concentrations of approximately 10^{10} EVs/ml [3]. In response to stimuli, such as cell activation due to inflammation and/or apoptosis, increased amounts of EVs are released. The frequencies of plasma EVs, which originate from different cellular origins, can be altered in disease states [4]. Over the last decade, the number of scientific publications describing physiological and pathological functions of EVs has increased significantly. The term"extracellular vesicles" is a collective term that encompasses various subtypes of cell-releasing membranous structures called exosomes, microvesicles, microparticles, ectosomes, oncosomes, apoptotic bodies, and many others. The International Society for Extracellular Vesicles (ISEV) proposed Minimal Information for Studies of Extracellular Vesicles

("MISEV") guidelines for accurate isolation and characterizations of EVs [5]. MISEV2018 proposes the classification of EVs according to their physical properties (size and density), biochemical composition (protein marker positivity), cells of origin or based on the description of the conditions that induce their release. The heterogeneity of EVs research is, apart from nomenclature, also a reflection of poorly standardized methods of isolation and downstream analysis. Complex biological samples containing non-EV contaminants pose a challenge for both the isolation and characterization of EVs. Usually a combination of different methods is used to obtain good data quality. The most common EVs are of platelet or megakaryocyte origin (> 50%) [6], while about 5-15% of EVs are of endothelial origin [7]. An increase in circulating EVs, especially endothelial EVs, is considered a hallmark of vascular dysfunction and cardiovascular disease. Increased EVs are found particularly in patients with hypertension [8], diabetes [9], acute coronary syndromes [10] and cardiovascular disease [11]. EVs, especially medium to large endothelial EVs, have been studied in patients with APS, who had significantly higher levels of circulating endothelial and platelet EVs compared with healthy controls [12]. One study also reported increased levels of small EVs (sEVs), which are less than 200 nm in size, in the plasma of patients with APS [13]. In addition, they reported on an altered protein profile of sEVs, indicating platelet and endothelial activation. These results show that a complex systemic network that exists in the form of cell–cell communication via sEVs is altered in APS patients.

2. Extracellular vesicles

Extracellular vesicles are small particles composed of a phospholipid bilayer that encloses soluble cytosolic or endosomal material and nuclear components and, unlike a cell, are unable to replicate. EVs can be as small as the smallest physically possible unilamellar liposome (about 20-30 nm) or as large as 1 μm or more [14]. EVs serve as regulators of the transfer of biological information (proteins, nucleic acids, lipids and metabolites), which act both locally and remotely [15]. EVs are found in a variety of human biofluids including serum, plasma, urine, saliva, breast milk, amniotic fluid, ascites fluid, cerebrospinal fluid and even bile [16]. Under normal physiological conditions, they are continuously secreted into the extracellular environment, however, the amount of EVs is increased by activated and apoptotic cells and is associated with different pathologies, including thrombosis [7]. EVs are probably the most extensively studied in cancer and were also found to play a significant role in cancer-associated thrombosis [17]. Over the last decade, EVs have been extensively studied in the field of biomedical research to determine their biological role in normal physiology and in disease state as well as to exploit potential clinical applications in the diagnosis and prognosis of disease. EVs are considered a promising source of biomarkers since they carry different biological materials that reflect the status of the cell of origin. Nevertheless, EVs have also been considered as a therapeutic agent, as an alternative to their synthetic counterparts, such as liposomes [18].

2.1 Classification of EVs

The classification and nomenclature of EVs is complicated and could be confusing due to overlapping definitions. The most common classification of EVs currently used in the literature is the classification of different EVs into subtypes, such as endosomal derived exosomes, membrane derived (microparticles, microvesicles or ectosomes) and apoptotic bodies. This classification is based on the assignment of a specific EV to a particular biogenesis pathway, which remains very difficult to assess [19]. Unless biogenesis is investigated directly, EVs are classified according to their a) physical characteristics such as size: "small EVs" (sEVs; size <100 nm or < 200 nm) and "medium/large" (m/lEVs; size >200 nm), and density; low, medium, high, with defined range, b) biochemical composition (surface expression or by the presence of a specific molecule within EVs), or c) description of a specific condition or cell of origin (**Figure 1**) [19].

2.2 Biological role of EVs

The key biological function of EVs is cell to cell communication and the transfer of biological materials that act closely, but also, and more importantly, remotely. Cargo within the EVs is protected from degradation in the bloodstream and can be successfully transferred to specific cells of interest, affecting several biological functions of these cells. EVs can transfer a wide variety of molecules: heat shock proteins (HSP-90, HSP-70), interleukins (IL), such as tumor necrosis factor-alpha (TNFα), acute phase proteins, such as serum amyloid A [20], enzymes, peptides, growth factors [14]. Therefore, EVs have a wide range of biological functions including immune response, antigen presentation, and the transfer of RNA, including micro RNA (miR) and DNA. Given the fact that EVs migrate through the bloodstream they can have pleiotropic effects that are likely to affect every tissue in the body [14]. In immunity, they modulate immune cells, cell–cell interactions, and transfer of cytokines and chemokines. In the heart and vessels, they stimulate

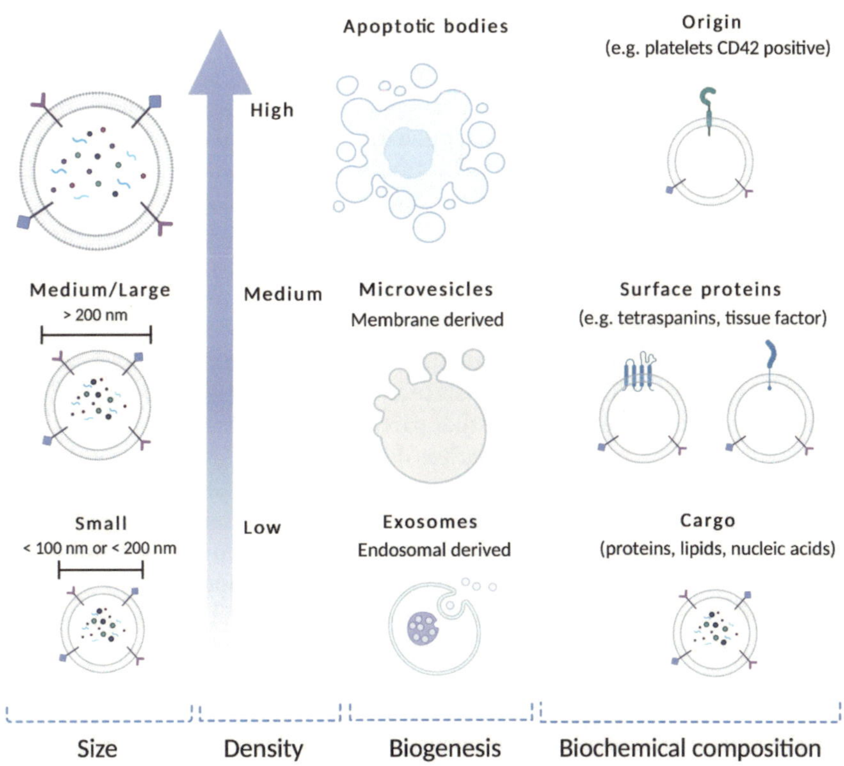

Figure 1.
Classification of EVs. EVs can be classified according to their size (Small <100 nm or < 200 nm, Medium/large >200 nm), density (Low, Medium, High) with a defined density range, biogenesis pathway (Exosomes; endosomal derived, Microvesicles; membrane derived and Apoptotic bodies; released upon cell apoptosis) or biochemical composition defining EVs origin, surface proteins or cargo. Created with BioRender.com.

coagulation and thrombosis, modulate angiogenesis, calcification and vascular repair. In the adipose tissue, they modulate angiogenesis, inflammation, cell differentiation and secretion of cytokines. In the bone marrow, they are involved in cell–cell cooperation, cell proliferation, differentiation and maturation. In the central nervous system, they are involved in the integration of neurons and various glial cells, modulate angiogenesis, neuronal plasticity and myelination. In the blood, they influence activation and aggregation of platelets, are directly involved in coagulation, as well as cargo transfer of procoagulant or anticoagulant molecules, cytokines and growth factors [14].

2.3 Methods for EVs isolation

Biological fluids containing EVs, which serve as potential minimally invasive liquid biopsies, have shifted its proteomic and genomic profiling research towards identification of biomarkers for disease diagnosis, prognosis and longitudinal monitoring. Studying EVs and their cargo typically requires separation from a biological matrix (such as a complex fluid or tissue) to analyze the unique EV components. However, isolating EVs from different sources presents certain challenges. For example, in serum and plasma the main challenge is to separate EVs from highly abundant non-EV proteins, such as albumin and globulins and non-EV lipid particles, such as lipoproteins and chylomicrons [21]. These co-purified contaminants pose a challenge for the isolation, analysis, and application of EVs. Correct interpretation and detailed reporting of the nature of EV samples and sample handling including storage, isolation, and analytical procedures for the analysis of EVs is required [18]. Many approaches have been used, including differential ultracentrifugation, density gradient ultracentrifugation, size exclusion chromatography, and affinity/immunoaffinity capture methods. All these approaches have their limitations and advantages, which are challenged by both the source and quantity of starting material and the downstream application [21]. Serial centrifugation enables the separation of EVs from cells, cell debris and larger vesicles such as apoptotic bodies. Ultracentrifugation (UC) exploits high centrifugal speed (100.000 x g) for a sufficient time to allow EVs to pellet. It separates particles based on their size, shape, and flotation density and is less efficient for smaller and less dense particles. Repeated centrifugation can reduce the amount of non-EVs particles, but also reduces the yield and may damage the EVs [21]. Density centrifugation or density ultracentrifugation uses a density gradient medium or cushion of denser solution (e.g. sucrose cushion; sUC) [22] to separate particles of a similar density. This technique takes advantage of the fact that particles denser than the solvent sediment in the suspension, while particles less dense float up. This increases the purity of samples and reduces the potential of mechanical damage to the vesicles [23]. Density gradient ultracentrifugation is successful in separating chylomicrons, very low-density, low-density and intermediate density lipoproteins present in plasma. However, particles of similar density, such as high-density lipoproteins, are co-isolated with the EVs [21]. Size exclusion chromatography (SEC) is a chromatographic method that allows vesicles of a particular size to be separated where EVs retain their structure and physiological function [24]. When performing SEC protein contaminants and aggregates of similar size, are often still present. In addition, the sample has to be further concentrated because of the different pooled fractions, decreasing the yield of isolation. Holcar et al. have investigated the purity of the samples by comparing sUC and SEC; the two most commonly used methods for the isolation of EVs. Transmission electron microscopy (TEM) of EVs isolated with SEC showed increased levels of lipoproteins. This was further confirmed by determining a significant increase of ApoA1 (found in high-density lipoproteins) and ApoB100 (found in very low-density, low-density and intermediate-density lipoproteins) [22]. Based on their results, the presence of lipoproteins in SEC isolates could have a significant impact on downstream analysis. Polymer-based precipitation uses volume-excluding polymers to lower the solubility of EVs and similarly sized non-EV particles which are isolated via low speed centrifugation. The main problem using this method is that protein removal kits must be used [21]. The highest purity of isolated EVs is achieved by using different immunopurification methods, such as immunomagnetic isolation. This method separates EVs on the basis of an antigen antibody interactions where the antibodies linked to the matrix (e.g. magnets) are directed against a specific antigen of interest on EVs [25]. Using this methodology, a specific EV subpopulation is investigated, however, the information about the general vesicle population is lost. In addition, when using the immunopurification method, EVs stay bound to the matrix, which makes them incompatible with certain downstream analyses (**Table 1**).

2.4 Methods for EVs analysis

The analysis of EVs is greatly hampered by their heterogeneity (size, different populations etc.) and by the complex nature of any biological or clinical sample (the presence of non-EVs contaminants). The characteristics of EVs can be determined by biochemical analysis (immunoblotting, immunosorbent EV assays and flow cytometry) or with physical analysis (electron microscopy (EM), atomic force microscopy (AFM), dynamic light scattering (DLS), nanoparticle tracking analysis (NTA), tunable restrictive pulse sensing (tRPS) and flow cytometry) as well as novel, optical based, technologies (fluorescence-based techniques, surface plasmon resonance, interferometric imaging and electrochemical sensing) [18] (**Table 1**). Due to challenges in EVs analysis, a combination of different methods is very common.

2.4.1 Physical analysis

The physical analyses of EVs involve determining a size range, shape and concentration. The size of EVs can be determined directly by high-resolution imaging, or indirectly, by using optical or electrical readouts. Direct high-resolution imaging includes microscopy methods, such as EM or AFM, to obtain an accurate estimate of individual EVs in nanoscale resolution [18]. EM is used to determine the size and morphology of individual EVs. This method employs an electron beam instead of light to obtain high-resolution images in nanoscale. The most commonly used EM techniques are scanning (SEM) and transmission (TEM). Scanning electron microscopy image will explore the topography of the EVs surface. Since electrons pass through the sample in TEM, a 2D image of EVs will be obtained, which will

Type of EVs	Isolation of EVs	Characterization of EVs
Small	Ultracentrifugation +/− density gradient, SEC, polymer-based approaches, immunopurification	AFM, EM, ELISA, NTA, RPS, DLS, WB
Medium/Large	Centrifugation +/− density gradient	AFM, EM, NTA, IF, ELISA, flow cytometry

Abbreviations: AFM: atomic-force microscopy; DLS: dynamic light scattering; ELISA: enzyme-linked immunosorbent assay; EM: electron microscopy; IF: immunofluorescence microscopy; NTA: nanoparticle tracking analysis; RPS; restrictive pulse sensing; SEC: size exclusion chromatography: EM: transmission electron microscopy WB: western blotting.

Table 1.
Most commonly used methods for isolation and characterization of EVs.

also provide the information on the inner structure [26]. These electron microscopy methods require fixation or drying of the sample which complicates the translation of observed structures to the native morphology of the EVs. To avoid sample dehydration variations of electron microscopy techniques, such as cryogenic TEM, have been evolved [27]. In the AFM, an extremely sharp tip scans the area and its deflection is translated into topology information. It provides additional information about mechanical properties, such as stiffness and elasticity of the vesicles. In most cases, AFM is performed on dry, immobilized surfaces, which in turn may damage the EVs [28]. This can be prevented by analyzing EVs in a solution [29]. Indirect methods estimate the size and concentration based on the interaction of EVs with light (DLS and flow cytometry), their diffusion trajectories (NTA or their effect on the electrical current (tRPS). DLS is based on the analysis of temporal intensity fluctuation of laser light scattered by a dispersion of freely diffusing EVs. Unlike EM and AFM it measures the collective mobility (diffusion coefficient) of scattering EVs that are present in the measured volume. Flow cytometry is often used to analyze the number of cells and their biochemical composition. EVs are much smaller than cells and are usually not detected due to the low sensitivity of the method. However, adapted protocols have been developed to enable the analysis of EVs [30]. In flow cytometry, the flow of cells is hydrodynamically focused in a flow chamber and enables the illumination of a single cell by several different lasers. The forward light scatter on the cell will allow information on the cells' sizes while the side scatter will gave information on the granularity and composition [31]. Because the EVs are very small and have a low refractive index, flow cytometers can more accurately determine the EVs larger than 500 nm. Smaller EVs are detected in the background signal and collectively due to the swarm effect, which happens when multiple EVs are simultaneously and not separately illuminated by the laser, creating a swarm [32]. The recent advances in the field of flow cytometry enable to detect also populations as small as 100 nm [33]. NTA measures how fast a particle diffuses in a static solution due to the principle of Brownian particle motion. By analyzing its motion trajectories, it determines the size distribution of vesicles. tRPS is a technique that measures changes in electrical current as each particle passes through an adjustable nanopore [18]. The heterogeneity of the samples is a major problem with all indirect methods. Compared to direct methods the number of EVs that can be analyzed is typically higher, which allows a better estimate of the concentration. This is also due to the fact that these vesicles are in their original state. However, these methods are not able to provide information on the presence of contaminants, such as lipoproteins.

2.4.2 Biochemical analysis

The characterization of EVs to determine the surface markers, markers of origin and proteins they carry allows to infer the functional role of these vesicles in health and disease. Methods might be divided to more conventional ones; the immunoblotting assays or the methods that will employ the capture of the vesicle; immunosorbent methods. Immunoblotting methods are based on the lysis of a vesicle and the analysis of its contents either by direct spotting on a membrane (dot blot) or separation of proteins using SDS PAGE combined with western blotting, in which specific proteins of interest are determined with labeled antibodies. Immunoblotting methods are often used to determine the presence of EVs in a sample. These methods can also be used to determine the purity of samples [18]. Immunosorbent assays are based on the detection of EVs using specific antibodies directed against surface proteins of EVs. Derived from the classical enzyme linked immunosorbent protein assay (ELISA), EVs are captured on a solid surface coated with antibodies that are typically present on the EVs. EVs capture results in a strong enrichment. Analysis of EVs surface proteins is afterwards performed with antibodies directed to a protein of interest on the surface of the EVs. These detection antibodies are conjugated to an enzyme enabling the conversion of a fluorescent/colored substrate that can be quantified with a spectrophotometer [18].

3. Extracellular vesicles in vascular pathologies

The main cell types involved in vascular hemostasis are endothelial cells, platelets and monocytes. All these cells release EVs, which leads to a complex interplay between different vesicles and different cells. EVs are continuously released in low concentrations from the cells into the intercellular environment, but this is greatly increased during cellular activation and apoptosis. EVs transmit various biological information (in the form of proteins, lipids and nucleic acids). Travelling through the bloodstream, EVs serve as local or distant messengers that transmit information to a variety of cells and tissues. Hemostasis is a very strictly regulated process that maintains normal function of vasculature despite the presence of triggers, such as injury and/or infection. One of the consequences of an altered hemostatic balance is the formation of thrombi, a process in which EVs play an important role [15]. EVs coming from activated cells have been shown to have both procoagulant and proinflammatory effects. Procoagulant effects are related to the fact that some EVs contain anionic phospholipids, mainly phosphatidylserine (PS), on their surface, which contributes to the assembly and activation of the prothrombinase complexes, thus promoting thrombin formation [34]. However, not all EVs carry PS on their surface, suggesting the involvement of other mechanisms contributing to the procoagulant state [35], including other important coagulation factors, such as tissue factor (TF), Factor XII [36], and reduced activity of tissue factor pathway inhibitor (TFPI) and thrombomodulin on endothelial cells [37]. In addition, EVs also induce the expression of adhesion molecules; integrins and selectins on the recipient cells causing platelets, monocytes, and endothelial cells to interact more intensively with each another. Finally, EVs also contribute significantly to the proinflammatory state in the vascular microenvironment by delivering or inducing certain cytokines and chemokines and by transferring nucleic acids and lipids [38]. The effects that these EVs have on different cell types disrupt the normal functioning of the vascular system, leading to the development of different pathologies, including deep vein thrombosis or pulmonary embolism [7] and cardiovascular diseases (atherosclerosis [39], hypertension [8], myocardial infarction [40] and stroke [41]).

3.1 Platelet-derived EVs

EVs from activated platelets can have different effects on endothelial cells, monocytes and other platelets (**Figure 2A**). Namely, increased levels of intracellular adhesion molecule-1 (ICAM-1), a well-known activator of endothelium was observed on endothelial cells upon stimulation with platelet EVs [42, 43], an effect later ascribed to miR-320b transfer [42]. Increased expression of lymphocyte function-associated antigen-1 LFA-1 (CD11a/CD18) and macrophage antigen-1 Mac-1 (CD11b/CD18); both important in mediating monocyte-endothelium interactions, were observed on monocytes upon stimulation with platelet EVs. These effects are induced by the transfer of arachidonic acid from platelet EVs and appear to be dependent on the activation of protein kinase C [44]. Platelet EVs therefore significantly modulate adhesion of monocytes to endothelial cells. It has also been shown that platelet EVs increase the deposition of platelets on damaged arteries and increase platelet aggregation and adhesion to collagen [45]. By influencing

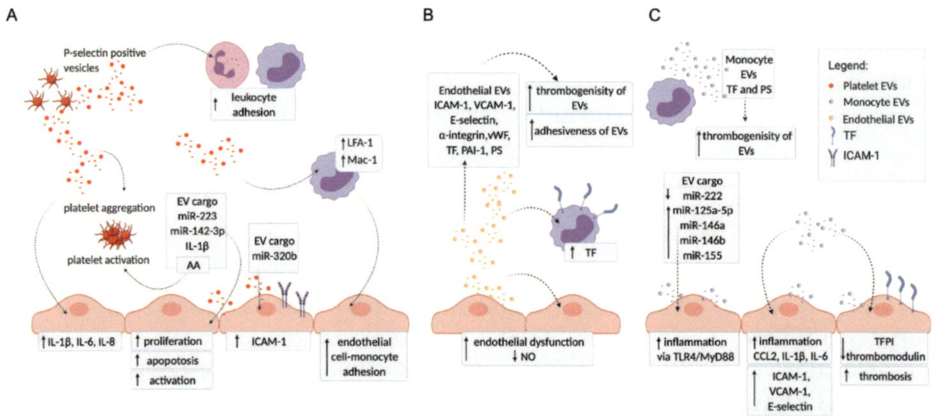

Figure 2.
Activation of platelets, monocytes and endothelial cells by EVs deriving from different cells. Schematic representation of the potential in vitro mechanisms focusing on vascular function, inflammation and thrombosis. (A) Platelet EVs can stimulate endothelial cells and monocytes via direct interaction or cargo delivery (miR and lipids). Furthermore, platelets EVs can also act via a feedback loop causing platelet aggregation and activation. Platelet EVs induce endothelial cell activation, proliferation and apoptosis by the transfer or miR-223 and miR-142-3p while ICAM-1 expression is induced by the delivery of miR-320b. Increased adhesion between endothelial cells and monocytes as well as between leukocytes in mediated by platelet EVs. (B) EVs released form endothelial cells were found to have a procoagulant profile expressing vWF, TF, PAI-1, PS as well as increased adhesive properties expressing VCAM-1, ICAM-1, E-selectin, and α-integrin. Endothelial EVs promote procoagulant profile of monocytes by induction of the TF expression on these cells. Endothelial EVs induce endothelial dysfunction by attenuating the production of nitric oxide from endothelial cells (C) Monocytes release procoagulant EVs that carry TF and PS. Furthermore, monocyte EVs interact with endothelial cells causing increased expression of adhesion molecules (ICAM-1, VCAM-1 and E-selectin), increased inflammation and procoagulant profile by reducing the expression of anticoagulant molecules (TFPI and Trombomodulin). Monocyte EVs transfer miR cargo (miR125a-5p, miR-222, miR-146a, miR-146b, miR-155) and induce inflammation in endothelial cells. CCL2, C-C motif chemokine ligand 2; ICAM-1, intercellular adhesion molecule 1; IL, interleukin; LFA1; lymphocyte function-associated antigen 1; Mac-1, Macrophage antigen-1; mIR; micro RNA; MyD88, myeloid differentiation primary response gene 88; NO, nitric oxide; PAI-1, plasminogen activator inhibitor-1; PS, phosphatidylserine; TF, tissue factor; TLR4, tool like receptor 4; VCAM-1, vascular cell adhesion molecule 1; vWF, von Willebrand factor. Created with BioRender.com.

cell adhesiveness, EVs also modulate interactions between leukocytes. Platelet EVs use P-selectin to bridge leukocytes, increase leukocyte-leukocyte interactions and enhance leukocyte accumulation on a P-selectin surface [46, 47]. Platelet EVs can therefore contribute to increased adhesion and aggregation of platelets and leukocytes on blood vessel walls during pathology. In addition, platelet EVs influence the production of cytokines (IL-1β, IL-6, IL-8) [43] and the transfer of miRNA (miRs 142-3p and 223), affecting the activation, proliferation and apoptosis of endothelial cells [48, 49]. In addition, platelet activation by the transfer of arachidonic acid from platelet EVs to other platelets, was observed [50]. Importantly, the role of platelet EVs in hemostasis is not entirely clear, as there is evidence that these EVs can also have anticoagulant effects [51, 52]. Further research is needed to determine, which key stimuli are responsible for determining the final effect of platelet EVs.

3.2 Endothelial-derived EVs

Endothelial cell activation and damage play an important role in vascular pathologies, with endothelial EVs being proposed as one of the causative agents in vascular pathologies (**Figure 2B**). Many proinflammatory factors (e.g. TNF-α, lipopolysaccharide, C-reactive protein and reactive oxygen species) and coagulation stimuli (thrombin, plasminogen activator inhibitor-1 (PAI-1)) can increase the levels of endothelial EVs. These vesicles carry adhesion molecules; ICAM-1, vascular cell adhesion protein 1 (VCAM-1), E-selectin, VE-cadherin, α-integrin, growth factors; endoglin, CD146, vascular endothelial growth factor (VEGF) receptor and molecules involved in coagulation, such as von Willebrand factor (vWF), TF, PAI-1 [53–55]. The expression of anionic phospholipids; such as PS, together with coagulation molecules, contribute to their procoagulant role. In addition, endothelial EVs may interact with other cells such as monocytes and induce the expression of TF on these cells [56]. Endothelial EVs induce endothelial dysfunction by attenuating the production of nitric oxide from endothelial cells [57]. Conversely, endothelial EVs may also have anticoagulant and antiinflammatory potential [38]. Although they exert different effects that are mostly dependent on the environment they originate from, endothelial EVs are generally believed to impair vascular function [58].

3.3 Monocyte-derived EVs

Leukocytes play an important role in the maintenance of vascular homeostasis. The activation of monocytes leads to increased release of monocyte EVs, which contribute to the disturbance of the hemostatic balance (**Figure 2C**). Monocyte EVs adhere to endothelial cells via LFA-1-ICAM-1 adhesion, as shown by the blocking of LFA-1 [37]. Once internalized, EVs were able to induce extracellular signal-regulated protein kinase (ERK1/2) and nuclear factor-κB (NF-κB) signaling pathways that increase the expression of the adhesion molecules VCAM-1, ICAM-1, and E-selectin on endothelial cells [59]. On the other hand, Tang et al. suggested that monocyte EVs induce *de novo* synthesis of ICAM-1, chemokine C-C motif ligand 2 (CCL2) and IL-1β in endothelial cells. This occurs via the activation of toll like receptor 4 (TLR4)/Myeloid differentiation primary response gene 88 (MyD88)/ NF-κB [60]. An increase in the adhesion profile of endothelial cells makes them more susceptible to interactions with platelets and monocytes and increase the prothrombotic state of the vasculature. Monocyte EVs trigger immune dysfunction related proinflammatory pathways also by the transfer of different miRs to the recipient cells. Levels of miR-125a-5p, miR-146a, miR-146b, miR-155 were significantly increased and miR-222 levels were decreased in INFα and lipopolysaccharide stimulated monocyte EVs compared to unstimulated monocyte EVs. Monocyte EVs transfer functional EVs to endothelial cells and activate the TLR4/MyD88/NF-κB signaling leading to differential expression of immunomodulatory miR in endothelial cells [61]. Both monocytes and monocyte EVs are positive for TF [37], a primary cellular initiator of blood coagulation. In vascular injury, TF forms a complex with factor VIIa, which activates the coagulation protease cascade and eventually leads to fibrin deposition and platelet activation [62]. In addition, monocyte EVs reduce the expression of the anticoagulant TFPI and of thrombomodulin on endothelial cells [37].

4. Pathological mechanisms of the Antiphospholipid syndrome

Antiphospholipid syndrome (APS) is a systemic autoimmune disorder characterized by venous and/or arterial thrombosis and pregnancy complications in the presence of antiphospholipid antibodies (aPL). aPL are a heterogeneous group of autoantibodies, of which anti-cardiolipin (anti-aCL), anti-β2 glycoprotein I (anti-β2GPI) and lupus anticoagulant (LA), are in the laboratory criteria for the diagnosis of APS [63]. In addition to criteria aPL other, non-criteria aPL, such as antibodies against phosphatidylserine/prothrombin complex, were found to play an important role in APS [64, 65]. These antibodies are, in some patients, the only elevated aPL. Although aPL are persistent in APS patients, thrombosis occurs only occasionally, suggesting the involvement of other triggers that, together with aPL, turn the hemostatic balance in favor of thrombosis. In the development of APS, a two hit theory has been proposed in which the continuous presence of aPL as the first hit and inflammation, trauma, or surgery as a second hit together lead to thrombus formation [66, 67]. APS pathogenesis clearly involves both inflammatory and coagulation pathways in endothelial cells, monocytes, neutrophils, and platelets. Frequently identified prothrombotic mechanism is inhibition of the natural anticoagulant pathways [68]. It has been shown that aPL inhibit the activation of protein C [69] and its ability to inactivate factors V and VIII [70]. In addition, aPL inhibit the activity of TFPI [71] and activation of antithrombin [72]. They have also been found to be involved in fibrinolysis by neutralizing the ability of anti-β2GPI to stimulate tissue-type plasminogen activator [73]. Furthermore, aPL impair the ability of Annexin A5 to form a network on procoagulant anionic phospholipids [74]. aPL also directly bind to vascular cells and trigger their activation, which in response, release prothrombotic molecules and thus contribute significantly to the pathogenesis of APS. The activation of endothelial cells leads to a disruption of the normally anticoagulant endothelial surface [68]. This is achieved by upregulating adhesion molecules (E-selectin, ICAM-1, VCAM-1) [75], molecules involved in coagulation (TF) [76] and by the decrease in endothelial cell derived nitric oxide [77]. The biochemical pathways are not fully defined, but research has suggested several receptor-mediated mechanisms including, annexin A2, TLR4/NF-κB, TLR2, TLR7 and low-density lipoprotein receptor-related protein 8 [68]. In addition to endothelial cells, aPL also act on platelets. Increased production of thromboxane B2, increased platelet adhesion to collagen type I and III and increased platelet activation have been described [66]. Among immune cells, monocytes are the most extensively studied in APS. In APS patients, monocytes have been shown to have a proinflammatory and procoagulant phenotype that is mediated by upregulation of NF-κB, MEK-1/ERK, and p38 MAP kinase pathways [78]. The main player of the procoagulant phenotype is increased surface expression, production and activity of TF on monocytes [79]. Stimulation of monocytes with aPL influences the release of IL-1β [80] and TNFα [81], probably by the activation of NLR family pyrin domain containing 3 inflammasome [82]. Monocyte-endothelial interactions are increased by upregulation of adhesion molecules on both cell types, as well as expression of other molecules, such as monocyte chemoattractant protein-1 by the endothelium, which in turn promotes the synthesis of TF by monocytes [83].

5. Extracellular vesicles in antiphospholipid syndrome: literature review and discussion

The role of EVs as communicators between different types of cells involved in the pathology of APS have been studied *in vivo* by analyzing the characteristics of EVs from plasma of APS patients and *in vitro* after stimulation of cells with aPL. As discussed above, EVs can carry characteristic proteins that determine their origin (**Figure 3**, upper panel) and their prothrombotic profile (e.g. by the presence of TF, PS) (**Figure 3**, lower panel). However, all EVs carry also different receptors, adhesion molecules and cargo (nucleic acids, lipids and proteins), which together influence the interaction between different cells, as well as information transfer. Larger vesicles (microvesicles) usually carry surface TF, PS and annexins while smaller EVs (exosomes) carry surface tetraspanins (CD9, CD63, CD81) and flotillin and alix, clathrin and TSG101 proteins as their cargo (**Figure 3**, lower panel).

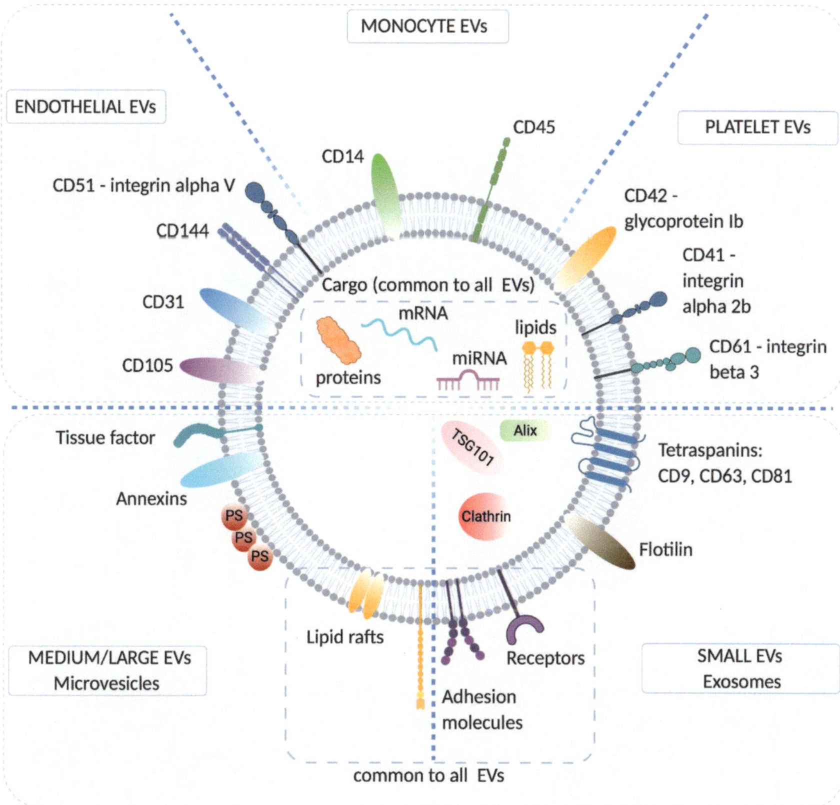

Figure 3.
Characterization of endothelial, monocyte and platelet EVs. Schematic representation of commonly expressed surface protein markers of endothelial cells, monocytes and platelets, as well as markers currently associated with small and medium/large EVs. Endothelial EVs usually express CD51 (Integrin alpha V) which is a part of a complex that binds extracellular matrix proteins, CD144 (Vascular endothelial cadherin), an important cell adhesion molecule in the formation of adherent junctions, CD31 (PECAM-1; platelet endothelial cell adhesion molecule) mediates leukocyte- and platelet-endothelial cell adhesion, CD105 (Endoglin) is a type I membrane glycoprotein and a part of transforming growth factor β receptor complex. Monocyte EVs commonly express CD14 (Cluster of differentiation 14) a known monocyte marker and CD45 (PTPRC; protein tyrosine phosphatase receptor type C) that is leukocyte specific cell surface glycoprotein involved in various cellular processes. Platelet EVs usually express different glycoproteins (CD42; glycoprotein IX, CD41; glycoprotein IIb, CD61; glycoprotein IIIa) that are integrin complex proteins involved in platelet aggregation. All EVs carry adhesion molecules, receptors and lipids that are involved in interaction of EVs with different cells. Furthermore, they carry proteins, nucleic acids and lipids that can be transferred to a target cell. Membrane derived vesicles-microvesicles, are usually larger and express procoagulant molecules, such as TF (Tissue factor), annexins and PS (Phosphatidylserine), whereas tetraspanins (CD9, CD63, CD81) and specific luminal proteins (Clathrin, TSG101 and Alix) are specific for smaller vesicles of endosomal origin-exosomes. Created with BioRender.com.

5.1 *In vivo* studies (characterization of EVs from plasma of APS patients)

The role of EVs has been studied in many vascular pathologies, including deep vein thrombosis [7] and cardiovascular disease [38], whose common denominator is endothelial dysfunction. In addition, platelet EVs have been proposed as a useful biomarker for long-term follow-up after myocardial infarction [84], whereas increases in the number of endothelial EVs play a role in many inflammatory diseases, such as atherosclerosis [39]. Studies investigating EVs in patients with APS are limited and heterogeneous (**Table 2**). To date and to our knowledge, there have been 13 studies investigating EVs in thrombotic APS patients. With one exception, all of them have focused on medium/large EVs. Furthermore, the results of these studies are not completely comparable because the methods for isolating

Reference	Patients	Controls	Isolation protocol	Method of quantification	Type of EVs	Main findings
Combes et al., 1999 [53]	5 APS, 8 APS + SLE	17 asympt. aPL+ (6 autoimmune, 4 infections, 5 malignancy, 2 undefined) 30 HBDs	2 x 1,500 × g (15″) 13,000 × g (1″)	AnxV+ or CD51+ < 1.5 μm (latex beads)	endothelial (CD51+)	↑ endothelial EVs in aPL+ pts. vs. HBDs. ↑ endothelial EVs in thrombotic aPL+ pts. vs. asympt. aPL+ pts. Levels of SLE aPL- pts. were similar to HBDs.
Joseph et al., 2001 [85]	20 APS 14 APS + SLE	16 SLE 20 HBDs	2 x 1,500 × g (15″) 13,000 × g (1″)	GPIIb-IIIa+ < 0.8 μm	platelet (GPIIb-IIIa+)	No difference in platelet EVs between APS pts., SLE pts. and HBDs.
Nagahama et al., 2003 [86]	24 APS 13 SLE + APS	30 HBDs	200 x g (10″, RT), 1000 x g (15″, RT)	AnxV+, CD42a+, CD14+	platelet (CD42a+) monocyte (AnxV+/CD14+)	↑ monocyte EVs in APS pts. vs. APS + SLE pts. and vs. HBDs. ↑ P-selectin+ platelets and platelet EVs in APS pts. vs. HBDs.
Dignat-George et al., 2004 [87]	23 APS 14 APS + SLE	28 SLE aPL+ no thrombosis 23 SLE aPL- no thrombosis 25 aPL- with thrombosis 25 HBDs	2 x 1,500 × g (15″) 13,000 × g (2″)	CD51+ < 0.8 μm (latex beads)	endothelial (CD51+)	↑ endothelial EVs in APS pts. vs. HBDs and vs. non aPL related thrombotic pts. ↑ endothelial EVs in SLE aPL+ pts. vs. HBDs. No difference between SLE aPL- pts. and non aPL related thrombotic pts. vs. HBDs. ↑ endothelial EVs in aPL+ pts. vs. aPL- pts. and vs. HBDs. No difference between primary or secondary APS.
Jy et al., 2007 [88]	60 APS	28 asympt. aPL+ 39 HBDs	160 × g (10″) 1500 × g (6″)	CD31+ or CD42+ < 1.5 μm	endothelial (CD31+/CD42-) platelet (CD31+/CD42+)	↑ platelet and endothelial EVs in APS pts. vs. HBDs. ↑ endothelial EVs in asympt. aPL+ pts. vs. HBDs. No difference in levels of endothelial EVs in APS pts. vs. asympt. aPL+ pts. ↑ platelet EVs in APS pts. vs. asympt. aPL+ pts. No difference in levels of platelet EVs in asympt. aPL+ pts. vs. HBDs.

Reference	Patients	Controls	Isolation protocol	Method of quantification	Type of EVs	Main findings
Flores-Nascimento et al., 2009 [89]	11 APS	9 DVT pts. at diagnosis 10 DVT pts. After 1-3 years of warfarin withdrawal 7 FVL pts. 37 HBDs	3000 x g (20″) 13,000 x g (30″)	AnxV+, CD14+, CD31+, CD45+, CD61+, CD142+, CD235+	total (AnxV+) platelet (CD61+) erythrocyte (CD235+) monocyte (CD14+) endothelial (CD31+) leukocyte (CD45+)	No difference in total EVs in DVT pts. at diagnosis, FVL pts., APS pts. and HBDs. ↑ total EVs in DVT 1-3 years and HBDs. No difference in platelet, erythrocyte, monocyte, endothelial and leukocyte EVs in all pts. groups vs. HBDs.
Vikerfors et al., 2012 [90]	40 APS, 12 secondary APS	52 HBDs	Isolation not described	phalloidin-, lacadherin+ or CD14+, CD42a+, CD142+, CD144+ < 1 μm (MegaMix beads)	total (lacadherin+) endothelial (CD144+) platelet (CD42a+) monocyte (CD14+) endothelial (CD144+/CD142+)	↑ total EVs in APS pts. vs. HBDs. ↑ endothelial, endothelial TF+ and monocyte EVs in APS pts. vs. HBDs. No difference in levels of platelet EVs in APS pts. vs. HBDs.
Willemze et al., 2014 [91]	11 APS 19 APS + SLE	72 asympt. aPL+	1,500 x g (10″, 4 °C) 2,000 x g (5″, 4°C) 20,000 x g (30″, 4°C)	not studied	TF+ EVs by a functional assay (TF activity)	↑ EV-TF activity in APS pts. vs. asympt. aPL+. No difference in EV-TF activity in the presence or absence of underlying SLE. No difference between different APS clinical complications. No correlation between EV-TF activity and aPL subtype.
Chaturvedi et al., 2015 [92]	47 aPL+ pts. (38 APS, 2 APS + SLE, 6 asympt. aPL+, 1 aPL+ migraine)	144 HBDs	2 x 1,500 × g (15″) 13,000 × g (2″)	AnxV+ or CD14+, CD41+, CD105+, CD142+, CD144+ < 1 μm (latex beads)	total (AnxV+) endothelial (CD105+/CD144+) platelet (CD41+) monocyte (CD14+) TF (CD142+)	↑ total EVs in aPL+ vs. HBDs. endothelial, platelet, and TF+ EVs in aPL+ vs. HBDs. No difference in levels of monocyte EVs in aPL+ vs. HBDs.

Reference	Patients	Controls	Isolation protocol	Method of quantification	Type of EVs	Main findings
Breen et al., 2015 [93]	66 aPL+ pts. (37 thrombotic APS, 11 obstetric APS, 18 asympt. aPL+).	18 HBD	2x 2,000 x g (15″, 4°C), 12,000 x g (2″, 4°C)	CD41+, CD51+, CD61+ or CD105+	endothelial (CD51+/ CD105+) platelet (CD41+/ CD61+)	↑ endothelial and platelet EVs in aPL+ pts. vs. HBDs. ↑ endothelial and platelet EVs in thrombotic APS pts. vs. HBDs. No difference in levels of endothelial and platelet EVs in obstetric APS pts. vs. HBDs. No difference in levels of endothelial and platelet EVs in asympt. aPL+ pts. vs. HBDs.
Niccolai et al., 2015 [94]	16 APS	16 asympt. aPL+ 16 HBDs	1,500 x g (15″) 3,000 x g (3″)	VPD450+ or CD31+, CD41a+, CD45+ < 0,9 μm (Megamix beads)	total (VPD450+ 7AAD-) endothelial (CD31+) platelet (CD41a+) leukocyte (CD45+)	↑ total, endothelial, platelet, and leukocyte EVs in APS pts. vs. HBDs, APS pts. vs. asympt. aPL+ pts. and asympt. aPL+ pts. vs. HBDs. ↑ total EVs in APS double and triple positivity vs. single positivity. Different EVs populations (endothelial, platelet and monocyte) did not correlate with aPL positivity. ↑ endothelial EVs in asympt. aPL+ pts. triple positivity vs. single positivity. Total, leukocyte and platelet EVs did not correlate with aPL positivity.
Hell et al., 2018 [95]	64 APS 18 APS + SLE 12 APS + LLD	30 HBDs	2,500 x g (15″, 15 °C)	not studied	TF+ EVs by a functional assay (TF activity)	No difference in EV-TF activity in LA+ pts. with thrombosis vs. HBDs. No difference in EV-TF activity in single, double or triple aPL+ patients. No difference in EV-TF activity in LA+ pts. with AT vs. VT vs. combination of both. No difference in EV-TF activity and the number of events (thromboses).

Reference	Patients	Controls	Isolation protocol	Method of quantification	Type of EVs	Main findings
Štok et al., 2020 [13]	14 APS	5 aPL- with thrombosis 7 HBD	820 x g (10″, RT) 2,500 x g (10″, RT) 10,000 x g, (45″, RT) 100,000 x g (2 h15", 4°C)	NTA	< 200 nm. Multiplex flow cytometry for 38 markers (detection via tetraspanins)	↑ sEVs in APS pts. vs. HBDs. Platelet (CD41b+, CD42a+), lymphocyte (CD8+), leukocyte (CD45+) and endothelial (CD31+) sEVs were detected. ↑ P-selectin on sEVs from APS pts. vs. HBDs. ↑ CD133/1 on sEVs from APS pts. vs. aPL- pts. with thrombosis.

Abbreviations: Anx V, annexin V; APS, antiphospholipid syndrome; AT, arterial thrombosis; aPL, antiphospholipid antibodies; asympt., asymptomatic; DVT, deep vein thrombosis; EVs, extracellular vesicles; FVL, factor V Leiden; HBDs, healthy blood donors; LLD, lupus like disease; NTA; nanoparticle tracking analysis; pts., patients; sEVs, small extracellular vesicles; SLE, systemic lupus erythematosus; TF, tissue factor; VT, venous thrombosis; ↑, elevated levels.

Table 2.
Isolation, quantification and characterization of EVs in plasma of APS patients.

and characterizing EVs are not standardized, the sample sizes in some studies are small and the patient population studied is very heterogeneous (e.g. patients with concomitant autoimmune or other disease). Overall, the studies investigated EVs from the three major cell types involved in the pathogenesis of APS: endothelium, platelets, and monocytes. Studies in the field of cardiovascular diseases and EVs have shown that both platelet and endothelial EVs are elevated in patients with hypertension, compared to healthy blood donors [8], therefore it is important to note that certain proportion of EVs detected in plasma of APS patients might be associated with hypertension. Correlations between the levels of EVs and systolic and diastolic blood pressure needs to be evaluated when investigating EVs in APS patients.

5.1.1 Medium and large extracellular vesicles

5.1.1.1 Endothelial-derived EVs

The endothelium is the major player in APS pathogenesis, so it is not surprising that endothelial EVs have been the most extensively studied (**Table 2**). Combes et al. published in 1999 the first study investigating endothelial EVs in APS using flow cytometry to detect endothelial marker integrin CD51+ EVs. They showed increased levels of endothelial EVs in LA+ patients compared to HBDs [53]. In addition, they have also showed a significant increase in endothelial EVs in LA+ patients with a history of thrombosis compared to asymptomatic LA+ patients. On the other hand, Jy et al. found no difference in endothelial EVs (CD31+/CD42-) between aPL+ thrombotic patients and asymptomatic aPL+ group, suggesting that the release of EVs might be related to the autoimmune process involving the presence of aPL [88]. Dignat-George et al. in 2004, showed increased levels of CD51+ endothelial EVs in APS patients and in aPL+ SLE patients compared to HBDs [87]. Increased levels of endothelial EVs were observed in aPL+ patients vs. HBDs as well as in aPL+ patients vs. aPL- patients. Increased levels of endothelial EVs in the plasma of APS patients compared to HBDs were later confirmed also in several other studies [90, 93, 94] (**Table 2**), in which different endothelial surface markers (CD31+, CD51+, CD105+, CD144+) were examined. Levels of endothelial EVs were shown to be increased in APS patients with exception of one study where the increase was not observed [89]. Chaturvedi et al., on the other hand investigated levels of TF+ endothelial EVs, and found them to be elevated in aPL+ patients, compared to HBDs [92]. A higher TF activity was also observed when comparing APS patients with asymptomatic aPL+ patients [91]. Contrarily, Hell et al. could not observe increased TF activity of endothelial EVs in APS patients vs. HBDs.

5.1.1.2 Monocyte- and Platelet-derived EVs

Platelet-derived EVs are the most numerous type of vesicles found in the circulation of healthy individuals [96], and their levels are further increased in disease [38]. They are known to play key roles in coagulation, thrombosis, vascular senescence and permeability. It has been suggested that platelet EVs induce vascular dysfunction and influence immune modulation, leading to vascular remodeling. Monocytes contribute to APS pathogenesis also by being the main source of tissue factor, which is one of the key initiators of the coagulation cascade. Similar to platelet EVs, it has been suggested that monocyte EVs cooperate in coagulation and vascular inflammation [38]. However, in APS, monocyte EVs (**Table 2**) have been less extensively studied compared to endothelial EVs. Joseph et al., showed no difference in plasma levels of

CD41+ platelet EVs between APS patients and HBDs [85]. This is consistent with the study by Vikenfors et al. (CD42a+) [97] and by Nascimento et al. (CD61+) [89]. On the other hand, increased levels of platelet EVs (CD41+, CD41a+, CD42+, CD42a+) were found in five other studies [86, 88, 92–94]. Jy et al. have shown an increase in platelet EVs in APS patients vs. asymptomatic aPL+ suggesting thrombosis rather than aPL may play a role in platelet EVs release [88]. An increase in monocyte EVs in APS patients compared to HBDs was observed by Nagahama et al. and Vikenfors et al. which is in contrast to two other studies where the authors could not see an increase [89, 92]. There is no consensus on whether platelet and monocyte EVs are elevated in APS patients and there is too little data to conclude on the effects of these EVs in APS patients.

5.1.2 Small extracellular vesicles

To date, only a study by Stok et al. has investigated the presence of sEVs in plasma of APS patients (**Table 2**). Compared to HBDs, significantly increased levels of sEVs were observed in APS patients. In addition, sEVs from different cellular origin: platelet (CD41b+, CD42a+), lymphocyte (CD8+), leukocyte (CD45+) and endothelial (CD31+) were detected. Flow cytometric characterization of sEVs defined a subpopulation of vesicles that were positive for P-selectin (CD62P) and the endothelial progenitor cell marker (CD133/1). sEVs from APS patients were enriched in surface expression of P-selectin, suggesting endothelial and platelet activation in APS. In addition, APS patients showed increased CD133/1 expression compared to aPL- patients with thrombosis, suggesting endothelial damage in APS [13]. The authors of this study suggest that increased levels of sEVs with distinct biological properties circulate in patients with thrombotic APS.

5.2 *In vitro* studies (characterization of EVs released by aPL stimulated cells)

One mechanism by which aPL promote thromboses is through their binding to endothelial cells causing the activation of endothelial cells [98, 99] which in response, release EVs that might modulate the activation of other adjacent cells [87, 100]. These effects were investigated on endothelial cells [87, 100–102] and placental explants [103] involving both small EVs and medium/large EVs (**Table 3**). A study by Dignat-George et al., showed a significant 4-fold increase in endothelial EVs with procoagulant activity after stimulation of human umbilical vein endothelial cells (HUVEC) with plasma of APS patients [87]. Only a moderate, non-significant increase was observed after HUVEC stimulation with the plasma from HBDs. In addition, endothelial EVs released after HUVEC stimulation with APS plasma, significantly reduced the normalized clotting time ratio. Wu et al. showed data where stimulation of HUVEC with anti-β2GPI caused the formation of an endothelial cell inflammasome and the release of EVs that were enriched in mature IL-1β, with a distinct mIR profile and caused endothelial activation [101]. However, activation of HUVEC does not appear to involve IL-1β receptor, but most likely follows the TLR/myd88-IRAK4 signaling pathway. Pericleous et al. [102] investigated the effect of purified polyclonal IgG from patients with APS (APS-IgG) and healthy controls (HC-IgG) on HUVEC [102]. HUVEC exposed to APS-IgG, produced significantly more endothelial EVs than those exposed to HC-IgG and a larger proportion of these EVs carried surface E-selectin. Levels of ICAM-1+, endoglin+ and VE-cadherin+ EVs did not differ from the ones stimulated with HC-IgG. VCAM-1+ and TF+ endothelial EVs could not be detected. Later Betapudi et al., also observed a 2-fold increase in levels of endothelial EVs released from HUVEC stimulated with anti-β2GPI [100]. EVs in obstetric APS patients

Reference	Cell type	Stimulation	Isolation protocol	Method of quantification of EVs	Levels	Other major findings
Dignat-George et al., 2004 [87]	HUVEC	plasma from APS pts. or HBDs	direct use of cell culture supernatant	AnxV+ (total EVs) < 0.8 μm (latex beads)	↑	↑ endothelial EVs with procoagulant activity.
Wu et al., 2015 [101]	HUVEC	anti-β2GPI purified from APS pts. plasma and from rabbits immunized with β2GPI	2,500 × g (15 “) 13,000 × g (2 “) 100,000 × g (90″)	qPCR, immunoblotting, inflammasome staining	NA	Anti-β2GPI caused formation of an endothelial cell inflammasome and the release of EVs that were enriched in mature IL-1β, had a distinct miR profile, and caused endothelial activation.
Pericleous et al., 2013 [102]	HUVEC	purified IgG from APS pts. and HBDs	3,000 × g (5″) 12,000 × g (60″)	AnxV+ (total EVs) CD62E+ (E-selectin), CD106+ (VCAM-1), CD54+ (ICAM-1), CD142+ (TF), CD105+ (endoglin), CD144+ (VE-cadherin) < 1 μm (latex beads)	↑ AnxV+ and E-selectin+	No difference in levels of ICAM-1+, endoglin+, and VE-cadherin+ EVs after APS IgG stimulation vs. HBD IgG. VCAM-1+ and TF+ EVs could not be detected.
Betapudi et al., 2013 [100]	HUVEC	anti-β2GPI purified from 3 APS pts., HBDs and rabbits immunized with β2GPI	1,500 × g, (30″) 13,000 × g (2″)	CD144+ < 1 μm (latex beads)	↑	Anti-β2GPI antibodies stimulate endothelial EVs release via nonmuscle myosin motor protein-dependent pathway.
Tong et al., 2017 [103]	1st trimester human placenta explants, HMEC-1	murine monoclonal anti-β2GPI, purified IgG from 5 APS pts. and HBDs	2,000 × g, (5″) 20,000 × g (60″) 100,000 × g (60″)	NTA	Not increased	↑ mean and modal size of EV after aPL stimulation. ↑ of mtDNA in EVs after aPL stimulation. EVs from placental explants activated HMEC-1 through TLR-9 receptor signaling.

Abbreviations: Anx V, annexin V; APS, antiphospholipid syndrome; aPL, antiphospholipid antibodies; β2GPI, β2-glycoprotein I; EVs, extracellular vesicles; HBDs, healthy blood donors; HUVEC, human umbilical vein endothelial cells; HMEC-1, human dermal microvascular endothelial cells; ICAM-1, intercellular adhesion molecule 1; IgG, immunoglobulin G; IL, interleukin; mtDNA, mitochondrial DNA; NA, not applicable; NTA, nanoparticle tracking analysis; pts., patients; TF, tissue factor; TLR, toll-like receptor; VCAM-1, vascular cell adhesion molecule 1; ↑, elevated levels.

Table 3.
Isolation, quantification and characterization of EVs derived from endothelial cells after stimulation with aPL.

were studied by Tong et al. [103], whereby exposure of first trimester human placental explants to monoclonal anti-β2GPI and IgG fractions from five anti-β2GPI positive APS patients did not affect the number or size of EVs. However, an increase in levels of mitochondrial DNA was observed in these vesicles that activated endothelial cells through a TLR-9-mediated pathway. This is supporting the idea that EV-associated mitochondrial DNA could be pathological in pregnant women with aPL.

6. Conclusions

Extracellular vesicles are small phospholipid bilayer particles that carry various biologically active molecules, such as proteins, lipids and nucleic acids. Their key biological function is cell–cell communication and the transfer of cargo. EVs normally circulate in the bloodstream of healthy individuals, but their levels are elevated in various pathological conditions, including APS. The classification, isolation and characterization of EVs has been developing in an accelerated manner over the last 20 years. Nevertheless, terms such as exosomes and microparticles are still present in the literature, but it is important to note that this classification is based on biogenesis, which is rather difficult to assess. It is therefore more optimal to classify EVs based on their other characteristics, such as size, density, origin etc. Each isolation and characterization techniques have their advantages and disadvantages and influences the properties of the EVs studied. Choosing the best combination, albeit of different isolation techniques, along with the characterization of EVs, is of utmost importance to achieve good data quality. In addition, the limitations of the methods used in both isolation and characterization must be considered. In the rapidly developing field of EVs research, variations of existing methods, as well as new technologies, are emerging that enable more precise isolation and characterization of EVs. EVs from platelets, monocytes and endothelial cells play a crucial role in vascular dysfunction, which is a causal factor in the disturbance of hemostasis and the development of thrombosis. Platelet and monocyte EVs are involved in the increased adhesiveness of endothelial cells and the increased interaction of leukocytes with the endothelium. Platelet, monocyte and endothelial EVs carry procoagulant molecules, such as TF, and modulate the expression of coagulation molecules in endothelial cells. Research on EVs in APS is very heterogeneous, due to the lack of standardization of isolation and characterization methods, all of which limits solid findings and conclusions. In addition to the technological challenges, EVs in APS are difficult to study because of the puzzling nature of APS. It is a chronic disease with a complex clinical spectrum due to many different features and symptoms (e.g. hypertension, thrombocytopenia). Patients with APS receive lifelong treatment with anticoagulants, and the actual acute phase is practically impossible to monitor. However, in view of the data on EVs in APS, a trend towards elevated total endothelial and platelet EV levels can be observed, suggesting an activated endothelium, even in the absence of an acute event. The results of the study of sEVs suggest that smaller vesicle populations may also play a role in the pathogenesis of APS. It appears that in patients with APS, levels of sEVs and different medium/large EVs are elevated. Further research is needed to confirm this in a larger number of patients as well as determine their functionality in APS. Data on increased levels of endothelial EVs in APS is supported by *in vitro* studies showing elevated levels of endothelial EVs following stimulation of endothelial cells with aPL. Studies investigating the role of aPL in vesicular release and its effects on the original cells also suggest that both small and medium/large EVs may play an important role in endothelial dysfunction in APS. However, future studies are needed to obtain a clearer picture of the signaling pathways and key molecules involved in interactions of EVs with the target cells.

Conflict of interest

The authors declare no conflict of interest.

Author details

Ula Štok[1,2*], Saša Čučnik[1,2], Snežna Sodin-Šemrl[1,3] and Polona Žigon[1,3]

1 Department of Rheumatology, University Medical Centre Ljubljana, Ljubljana, Slovenia

2 Faculty of Pharmacy, University of Ljubljana, Ljubljana, Slovenia

3 Faculty of Mathematics, Natural Sciences and Information Technologies, University of Primorska, Koper, Slovenia

*Address all correspondence to: ulastok@gmail.com

IntechOpen

© 2021 The Author(s). Licensee IntechOpen. This chapter is distributed under the terms of the Creative Commons Attribution License (http://creativecommons.org/licenses/by/3.0), which permits unrestricted use, distribution, and reproduction in any medium, provided the original work is properly cited.

References

[1] Miyakis S, Lockshin MD, Atsumi T, Branch DW, Brey RL, Cervera R, et al. International consensus statement on an update of the classification criteria for definite antiphospholipid syndrome (APS). *Journal of thrombosis and haemostasis: JTH*. **2006**; 4 (2); 295-306.

[2] Zhang J, Li S, Li L, Li M, Guo C, Yao J, et al. Exosome and exosomal microRNA: trafficking, sorting, and function. *Genomics Proteomics Bioinformatics*. **2015**; 13 (1); 17-24.

[3] Johnsen KB, Gudbergsson JM, Andresen TL, Simonsen JB. What is the blood concentration of extracellular vesicles? Implications for the use of extracellular vesicles as blood-borne biomarkers of cancer. *Biochim Biophys Acta Rev Cancer*. **2019**; 1871 (1); 109-16.

[4] Chaturvedi S, Alluri R, McCrae KR. Extracellular Vesicles in the Antiphospholipid Syndrome. *Semin Thromb Hemost*. **2018**; 44 (5); 493-504.

[5] Théry C, Witwer KW. Minimal information for studies of extracellular vesicles 2018 (MISEV2018): a position statement of the International Society for Extracellular Vesicles and update of the MISEV2014 guidelines. **2018**; 7 (1); 1535750.

[6] Flaumenhaft R, Mairuhu AT, Italiano JE. Platelet- and megakaryocyte-derived microparticles. *Semin Thromb Hemost*. **2010**; 36 (8); 881-7.

[7] Zara M, Guidetti GF, Camera M, Canobbio I, Amadio P, Torti M, et al. Biology and Role of Extracellular Vesicles (EVs) in the Pathogenesis of Thrombosis. *Int J Mol Sci*. **2019**; 20 (11).

[8] Preston RA, Jy W, Jimenez JJ, Mauro LM, Horstman LL, Valle M, et al. Effects of severe hypertension on endothelial and platelet microparticles. *Hypertension (Dallas, Tex : 1979)*. **2003**; 41 (2); 211-7.

[9] Sabatier F, Darmon P, Hugel B, Combes V, Sanmarco M, Velut JG, et al. Type 1 and type 2 diabetic patients display different patterns of cellular microparticles. *Diabetes*. **2002**; 51 (9); 2840-5.

[10] Bernal-Mizrachi L, Jy W, Jimenez JJ, Pastor J, Mauro LM, Horstman LL, et al. High levels of circulating endothelial microparticles in patients with acute coronary syndromes. *American heart journal*. **2003**; 145 (6); 962-70.

[11] Jansen F, Nickenig G, Werner N. Extracellular Vesicles in Cardiovascular Disease. *Circulation research*. **2017**; 120 (10); 1649-57.

[12] Chaturvedi S, Alluri R, McCrae KR. Extracellular Vesicles in the Antiphospholipid Syndrome. *Seminars in thrombosis and hemostasis*. **2017**.

[13] Stok U, Blokar E, Lenassi M, Holcar M, Frank-Bertoncelj M, Erman A, et al. Characterization of Plasma-Derived Small Extracellular Vesicles Indicates Ongoing Endothelial and Platelet Activation in Patients with Thrombotic Antiphospholipid Syndrome. *Cells*. **2020**; 9 (5).

[14] Berezin AE, Berezin AA. Endothelial cell-derived extracellular vesicles in atherosclerosis: the emerging value for diagnosis, risk stratification and prognostication. *Vessel Plus*. **2020**; 2020.

[15] Boulanger CM, Loyer X, Rautou PE, Amabile N. Extracellular vesicles in coronary artery disease. *Nat Rev Cardiol*. **2017**; 14 (5); 259-72.

[16] Raposo G, Stoorvogel W. Extracellular vesicles: exosomes, microvesicles, and friends. *J Cell Biol*. **2013**; 200 (4); 373-83.

[17] Almeida VH, Rondon AMR, Gomes T, Monteiro RQ. Novel Aspects

of Extracellular Vesicles as Mediators of Cancer-Associated Thrombosis. *Cells*. **2019**; 8 (7).

[18] Hartjes TA, Mytnyk S, Jenster GW, van Steijn V, van Royen ME. Extracellular Vesicle Quantification and Characterization: Common Methods and Emerging Approaches. *Bioengineering (Basel)*. **2019**; 6 (1).

[19] Thery C, Witwer KW, Aikawa E, Alcaraz MJ, Anderson JD, Andriantsitohaina R, et al. Minimal information for studies of extracellular vesicles 2018 (MISEV2018): a position statement of the International Society for Extracellular Vesicles and update of the MISEV2014 guidelines. *J Extracell Vesicles*. **2018**; 7 (1); 1535750.

[20] Schneider A, Simons M. Exosomes: vesicular carriers for intercellular communication in neurodegenerative disorders. *Cell Tissue Res*. **2013**; 352 (1); 33-47.

[21] Brennan K, Martin K, FitzGerald SP, O'Sullivan J, Wu Y, Blanco A, et al. A comparison of methods for the isolation and separation of extracellular vesicles from protein and lipid particles in human serum. *Sci Rep*. **2020**; 10 (1); 1039.

[22] Holcar M, Ferdin J, Sitar S, Tušek-Žnidarič M, Dolžan V, Plemenitaš A, et al. Enrichment of plasma extracellular vesicles for reliable quantification of their size and concentration for biomarker discovery. *Scientific Reports*. **2020**; 10 (1).

[23] Puzar Dominkus P, Stenovec M, Sitar S, Lasic E, Zorec R, Plemenitas A, et al. PKH26 labeling of extracellular vesicles: Characterization and cellular internalization of contaminating PKH26 nanoparticles. *Biochim Biophys Acta Biomembr*. **2018**; 1860 (6); 1350-61.

[24] Gámez-Valero A, Monguió-Tortajada M, Carreras-Planella L, Franquesa M, Beyer K, Borràs FE. Size-Exclusion Chromatography-based isolation minimally alters Extracellular Vesicles' characteristics compared to precipitating agents. *Sci Rep*. **2016**; 6; 33641.

[25] Koliha N, Wiencek Y, Heider U, Jungst C, Kladt N, Krauthauser S, et al. A novel multiplex bead-based platform highlights the diversity of extracellular vesicles. *J Extracell Vesicles*. **2016**; 5; 29975.

[26] Bozzola JJ RL. Electron Microscopy. Second Edition ed. MA, USA: Jones and Bartlett 1999.

[27] Cizmar P, Yuana Y. Detection and Characterization of Extracellular Vesicles by Transmission and Cryo-Transmission Electron Microscopy. *Methods Mol Biol*. **2017**; 1660; 221-32.

[28] Allison DP, Mortensen NP, Sullivan CJ, Doktycz MJ. Atomic force microscopy of biological samples. *Wiley Interdiscip Rev Nanomed Nanobiotechnol*. **2010**; 2 (6); 618-34.

[29] Sebaihi N, De Boeck B, Yuana Y, Nieuwland R, Pétry J. Dimensional characterization of extracellular vesicles using atomic force microscopy. *Measurement Science and Technology*. **2017**; 28 (3).

[30] Poncelet P, Robert S, Bailly N, Garnache-Ottou F, Bouriche T, Devalet B, et al. Tips and tricks for flow cytometry-based analysis and counting of microparticles. *Transfus Apher Sci*. **2015**; 53 (2); 110-26.

[31] Adan A, Alizada G, Kiraz Y, Baran Y, Nalbant A. Flow cytometry: basic principles and applications. *Crit Rev Biotechnol*. **2017**; 37 (2); 163-76.

[32] van der Pol E, van Gemert MJ, Sturk A, Nieuwland R, van Leeuwen TG. Single vs. swarm detection of microparticles and exosomes by flow cytometry. *J Thromb Haemost*. **2012**; 10 (5); 919-30.

[33] Stoner SA, Duggan E, Condello D, Guerrero A, Turk JR, Narayanan PK, et al. High sensitivity flow cytometry of membrane vesicles. *Cytometry A*. **2016**; 89 (2); 196-206.

[34] Owens AP, 3rd, Mackman N. Microparticles in hemostasis and thrombosis. *Circ Res*. **2011**; 108 (10); 1284-97.

[35] Perez-Pujol S, Marker PH, Key NS. Platelet microparticles are heterogeneous and highly dependent on the activation mechanism: studies using a new digital flow cytometer. *Cytometry A*. **2007**; 71 (1); 38-45.

[36] Van Der Meijden PE, Van Schilfgaarde M, Van Oerle R, Renné T, ten Cate H, Spronk HM. Platelet- and erythrocyte-derived microparticles trigger thrombin generation via factor XIIa. *J Thromb Haemost*. **2012**; 10 (7); 1355-62.

[37] Aharon A, Tamari T, Brenner B. Monocyte-derived microparticles and exosomes induce procoagulant and apoptotic effects on endothelial cells. *Thromb Haemost*. **2008**; 100 (5); 878-85.

[38] Oggero S, Austin-Williams S, Norling LV. The Contrasting Role of Extracellular Vesicles in Vascular Inflammation and Tissue Repair. *Front Pharmacol*. **2019**; 10; 1479.

[39] Charla E, Mercer J, Maffia P, Nicklin SA. Extracellular vesicle signalling in atherosclerosis. *Cell Signal*. **2020**; 75; 109751.

[40] Zhang Y, Cheng J, Chen F, Wu C, Zhang J, Ren X, et al. Circulating endothelial microparticles and miR-92a in acute myocardial infarction. *Biosci Rep*. **2017**; 37 (2).

[41] Chen Y, Xiao Y, Lin Z, Xiao X, He C, Bihl JC, et al. The Role of Circulating Platelets Microparticles and Platelet Parameters in Acute Ischemic Stroke Patients. *J Stroke Cerebrovasc Dis*. **2015**; 24 (10); 2313-20.

[42] Gidlöf O, van der Brug M, Ohman J, Gilje P, Olde B, Wahlestedt C, et al. Platelets activated during myocardial infarction release functional miRNA, which can be taken up by endothelial cells and regulate ICAM1 expression. *Blood*. **2013**; 121 (19); 3908-17, s1-26.

[43] Nomura S, Tandon NN, Nakamura T, Cone J, Fukuhara S, Kambayashi J. High-shear-stress-induced activation of platelets and microparticles enhances expression of cell adhesion molecules in THP-1 and endothelial cells. *Atherosclerosis*. **2001**; 158 (2); 277-87.

[44] Barry OP, Praticò D, Savani RC, FitzGerald GA. Modulation of monocyte-endothelial cell interactions by platelet microparticles. *J Clin Invest*. **1998**; 102 (1); 136-44.

[45] Suades R, Padró T, Vilahur G, Badimon L. Circulating and platelet-derived microparticles in human blood enhance thrombosis on atherosclerotic plaques. *Thromb Haemost*. **2012**; 108 (6); 1208-19.

[46] Forlow SB, McEver RP, Nollert MU. Leukocyte-leukocyte interactions mediated by platelet microparticles under flow. *Blood*. **2000**; 95 (4); 1317-23.

[47] Kuravi SJ, Harrison P, Rainger GE, Nash GB. Ability of Platelet-Derived Extracellular Vesicles to Promote Neutrophil-Endothelial Cell Interactions. *Inflammation*. **2019**; 42 (1); 290-305.

[48] Bao H, Chen YX, Huang K, Zhuang F, Bao M, Han Y, et al. Platelet-derived microparticles promote endothelial cell proliferation in hypertension via miR-142-3p. *Faseb j*. **2018**; 32 (7); 3912-23.

[49] Pan Y, Liang H, Liu H, Li D, Chen X, Li L, et al. Platelet-secreted microRNA-223 promotes endothelial

cell apoptosis induced by advanced glycation end products via targeting the insulin-like growth factor 1 receptor. *J Immunol.* **2014**; 192 (1); 437-46.

[50] Barry OP, Pratico D, Lawson JA, FitzGerald GA. Transcellular activation of platelets and endothelial cells by bioactive lipids in platelet micro-particles. *J Clin Invest.* **1997**; 99 (9); 2118-27.

[51] Somajo S, Koshiar RL, Norström E, Dahlbäck B. Protein S and factor V in regulation of coagulation on platelet microparticles by activated protein C. *Thromb Res.* **2014**; 134 (1); 144-52.

[52] Srikanthan S, Li W, Silverstein RL, McIntyre TM. Exosome poly-ubiquitin inhibits platelet activation, down-regulates CD36 and inhibits pro-atherothombotic cellular functions. *J Thromb Haemost.* **2014**; 12 (11); 1906-17.

[53] Combes V, Simon AC, Grau GE, Arnoux D, Camoin L, Sabatier F, et al. In vitro generation of endothelial microparticles and possible pro-thrombotic activity in patients with lupus anticoagulant. *J Clin Invest.* **1999**; 104 (1); 93-102.

[54] Abid Hussein MN, Böing AN, Biró E, Hoek FJ, Vogel GM, Meuleman DG, et al. Phospholipid composition of in vitro endothelial microparticles and their in vivo thrombogenic properties. *Thromb Res.* **2008**; 121 (6); 865-71.

[55] Dignat-George F, Boulanger CM. The many faces of endothelial microparticles. *Arterioscler Thromb Vasc Biol.* **2011**; 31 (1); 27-33.

[56] Sabatier F, Roux V, Anfosso F, Camoin L, Sampol J, Dignat-George F. Interaction of endothelial microparticles with monocytic cells in vitro induces tissue factor-dependent procoagulant activity. *Blood.* **2002**; 99 (11); 3962-70.

[57] Brodsky SV, Zhang F, Nasjletti A, Goligorsky MS. Endothelium-derived microparticles impair endothelial function in vitro. *Am J Physiol Heart Circ Physiol.* **2004**; 286 (5); H1910-5.

[58] Hromada C, Muhleder S, Grillari J, Redl H, Holnthoner W. Endothelial Extracellular Vesicles-Promises and Challenges. *Front Physiol.* **2017**; 8; 275.

[59] Wang JG, Williams JC, Davis BK, Jacobson K, Doerschuk CM, Ting JP, et al. Monocytic microparticles activate endothelial cells in an IL-1beta-dependent manner. *Blood.* **2011**; 118 (8); 2366-74.

[60] Tang N, Sun B, Gupta A, Rempel H, Pulliam L. Monocyte exosomes induce adhesion molecules and cytokines via activation of NF-kappaB in endothelial cells. *FASEB J.* **2016**; 30 (9); 3097-106.

[61] Dalvi P, Sun B, Tang N, Pulliam L. Immune activated monocyte exosomes alter microRNAs in brain endothelial cells and initiate an inflammatory response through the TLR4/MyD88 pathway. *Sci Rep.* **2017**; 7 (1); 9954.

[62] Mackman N. Role of tissue factor in hemostasis, thrombosis, and vascular development. *Arterioscler Thromb Vasc Biol.* **2004**; 24 (6); 1015-22.

[63] Radic M, Pattanaik D. Cellular and Molecular Mechanisms of Anti-Phospholipid Syndrome. *Front Immunol.* **2018**; 9; 969.

[64] Žigon P, Čučnik S, Ambrožič A, Kveder T, Šemrl SS, Rozman B, et al. Detection of antiphosphatidylserine/prothrombin antibodies and their potential diagnostic value. *Clin Dev Immunol.* **2013**; 2013; 724592.

[65] Žigon P, Podovšovnik A, Ambrožič A, Tomšič M, Hočevar A, Gašperšič N, et al. Added value of non-criteria antiphospholipid antibodies for antiphospholipid syndrome: lessons learned from year-long routine measurements. *Clin Rheumatol.* **2019**; 38 (2); 371-8.

[66] Corban MT, Duarte-Garcia A, McBane RD, Matteson EL, Lerman LO, Lerman A. Antiphospholipid Syndrome: Role of Vascular Endothelial Cells and Implications for Risk Stratification and Targeted Therapeutics. *J Am Coll Cardiol*. **2017**; 69 (18); 2317-30.

[67] Artenjak A, Omersel J, Ahlin Grabnar P, Mlinarič-Raščan I, Shoenfeld Y, Sodin-Semrl S, et al. Oxidatively altered IgG with increased immunoreactivity to β2-glycoprotein I and its peptide clusters influence human coronary artery endothelial cells. *Lupus*. **2015**; 24 (4-5); 448-62.

[68] Chaturvedi S, McCrae KR. Clinical Risk Assessment in the Antiphospholipid Syndrome: Current Landscape and Emerging Biomarkers. *Curr Rheumatol Rep*. **2017**; 19 (7); 43.

[69] Wahl D, Membre A, Perret-Guillaume C, Regnault V, Lecompte T. Mechanisms of antiphospholipid-induced thrombosis: effects on the protein C system. *Curr Rheumatol Rep*. **2009**; 11 (1); 77-81.

[70] Borrell M, Sala N, de Castellarnau C, Lopez S, Gari M, Fontcuberta J. Immunoglobulin fractions isolated from patients with antiphospholipid antibodies prevent the inactivation of factor Va by activated protein C on human endothelial cells. *Thromb Haemost*. **1992**; 68 (3); 268-72.

[71] Liestøl S, Sandset PM, Jacobsen EM, Mowinckel MC, Wisløff F. Decreased anticoagulant response to tissue factor pathway inhibitor type 1 in plasmas from patients with lupus anticoagulants. *Br J Haematol*. **2007**; 136 (1); 131-7.

[72] Shibata S, Harpel PC, Gharavi A, Rand J, Fillit H. Autoantibodies to heparin from patients with antiphospholipid antibody syndrome inhibit formation of antithrombin III-thrombin complexes. *Blood*. **1994**; 83 (9); 2532-40.

[73] Bu C, Gao L, Xie W, Zhang J, He Y, Cai G, et al. beta2-glycoprotein i is a cofactor for tissue plasminogen activator-mediated plasminogen activation. *Arthritis Rheum*. **2009**; 60 (2); 559-68.

[74] Rand JH, Wu XX, Quinn AS, Chen PP, McCrae KR, Bovill EG, et al. Human monoclonal antiphospholipid antibodies disrupt the annexin A5 anticoagulant crystal shield on phospholipid bilayers: evidence from atomic force microscopy and functional assay. *Am J Pathol*. **2003**; 163 (3); 1193-200.

[75] Simantov R, LaSala JM, Lo SK, Gharavi AE, Sammaritano LR, Salmon JE, et al. Activation of cultured vascular endothelial cells by anti-phospholipid antibodies. *J Clin Invest*. **1995**; 96 (5); 2211-9.

[76] Branch DW, Rodgers GM. Induction of endothelial cell tissue factor activity by sera from patients with antiphospholipid syndrome: a possible mechanism of thrombosis. *Am J Obstet Gynecol*. **1993**; 168 (1 Pt 1); 206-10.

[77] Mineo C. Inhibition of nitric oxide and antiphospholipid antibody-mediated thrombosis. *Curr Rheumatol Rep*. **2013**; 15 (5); 324.

[78] López-Pedrera C, Buendía P, Cuadrado MJ, Siendones E, Aguirre MA, Barbarroja N, et al. Antiphospholipid antibodies from patients with the antiphospholipid syndrome induce monocyte tissue factor expression through the simultaneous activation of NF-kappaB/Rel proteins via the p38 mitogen-activated protein kinase pathway, and of the MEK-1/ERK pathway. *Arthritis Rheum*. **2006**; 54 (1); 301-11.

[79] van den Hoogen LL, van Roon JA, Radstake TR, Fritsch-Stork RD, Derksen RH. Delineating the deranged immune system in the antiphospholipid syndrome. *Autoimmun Rev*. **2016**; 15 (1); 50-60.

[80] Hurst J, Prinz N, Lorenz M, Bauer S, Chapman J, Lackner KJ, et al. TLR7 and TLR8 ligands and antiphospholipid antibodies show synergistic effects on the induction of IL-1beta and caspase-1 in monocytes and dendritic cells. *Immunobiology*. **2009**; 214 (8); 683-91.

[81] Döring Y, Hurst J, Lorenz M, Prinz N, Clemens N, Drechsler MD, et al. Human antiphospholipid antibodies induce TNFalpha in monocytes via Toll-like receptor 8. *Immunobiology*. **2010**; 215 (3); 230-41.

[82] Müller-Calleja N, Köhler A, Siebald B, Canisius A, Orning C, Radsak M, et al. Cofactor-independent antiphospholipid antibodies activate the NLRP3-inflammasome via endosomal NADPH-oxidase: implications for the antiphospholipid syndrome. *Thromb Haemost*. **2015**; 113 (5); 1071-83.

[83] Velasquez M, Rojas M, Abrahams VM, Escudero C, Cadavid AP. Mechanisms of Endothelial Dysfunction in Antiphospholipid Syndrome: Association With Clinical Manifestations. *Front Physiol*. **2018**; 9; 1840.

[84] Stępień E, Stankiewicz E, Zalewski J, Godlewski J, Zmudka K, Wybrańska I. Number of microparticles generated during acute myocardial infarction and stable angina correlates with platelet activation. *Arch Med Res*. **2012**; 43 (1); 31-5.

[85] Joseph JE, Harrison P, Mackie IJ, Isenberg DA, Machin SJ. Increased circulating platelet-leucocyte complexes and platelet activation in patients with antiphospholipid syndrome, systemic lupus erythematosus and rheumatoid arthritis. *Br J Haematol*. **2001**; 115 (2); 451-9.

[86] Nagahama M, Nomura S, Kanazawa S, Ozaki Y, Kagawa H, Fukuhara S. Significance of anti-oxidized LDL antibody and monocyte-derived microparticles in anti-phospholipid antibody syndrome. *Autoimmunity*. **2003**; 36 (3); 125-31.

[87] Dignat-George F, Camoin-Jau L, Sabatier F, Arnoux D, Anfosso F, Bardin N, et al. Endothelial microparticles: a potential contribution to the thrombotic complications of the antiphospholipid syndrome. *Thromb Haemost*. **2004**; 91 (4); 667-73.

[88] Jy W, Tiede M, Bidot CJ, Horstman LL, Jimenez JJ, Chirinos J, et al. Platelet activation rather than endothelial injury identifies risk of thrombosis in subjects positive for antiphospholipid antibodies. *Thromb Res*. **2007**; 121 (3); 319-25.

[89] Flores-Nascimento MC, Beltrame MP, De Paula EV, Montalvao SL, Pereira FG, Orsi FL, et al. Microparticles in deep venous thrombosis, antiphospholipid syndrome and Factor V Leiden. *Platelets*. **2009**; 20 (6); 367-75.

[90] Vikerfors A, Mobarrez F, Bremme K, Holmström M, Ågren A, Eelde A, et al. Studies of microparticles in patients with the antiphospholipid syndrome (APS). *Lupus*. **2012**; 21 (7); 802-5.

[91] Willemze R, Bradford RL, Mooberry MJ, Roubey RA, Key NS. Plasma microparticle tissue factor activity in patients with antiphospholipid antibodies with and without clinical complications. *Thromb Res*. **2014**; 133 (2); 187-9.

[92] Chaturvedi S, Cockrell E, Espinola R, Hsi L, Fulton S, Khan M, et al. Circulating microparticles in patients with antiphospholipid antibodies: characterization and associations. *Thromb Res*. **2015**; 135 (1); 102-8.

[93] Breen KA, Sanchez K, Kirkman N, Seed PT, Parmar K, Moore GW, et al. Endothelial and platelet microparticles in patients with antiphospholipid antibodies. *Thromb Res*. **2015**; 135 (2); 368-74.

[94] Niccolai E, Squatrito D, Emmi G, Silvestri E, Emmi L, Ciucciarelli L, et al. A new cytofluorimetric approach to evaluate the circulating microparticles in subjects with antiphospholipid antibodies. *Thromb Res*. **2015**; 136 (6); 1252-8.

[95] Hell L, Ay C, Posch F, Gebhart J, Koder S, Mackman N, et al. Low extracellular vesicle-associated tissue factor activity in patients with persistent lupus anticoagulant and a history of thrombosis. *Ann Hematol*. **2019**; 98 (2); 313-9.

[96] Lynch SF, Ludlam CA. Plasma microparticles and vascular disorders. *Br J Haematol*. **2007**; 137 (1); 36-48.

[97] Vikerfors A, Mobarrez F, Bremme K, Holmstrom M, Agren A, Eelde A, et al. Studies of microparticles in patients with the antiphospholipid syndrome (APS). *Lupus*. **2012**; 21 (7); 802-5.

[98] Velásquez M, Rojas M, Abrahams VM, Escudero C, Cadavid ÁP. Mechanisms of Endothelial Dysfunction in Antiphospholipid Syndrome: Association With Clinical Manifestations. *Frontiers in Physiology*. **2018**; 9.

[99] Lackner KJ, Muller-Calleja N. Pathogenesis of antiphospholipid syndrome: recent insights and emerging concepts. *Expert review of clinical immunology*. **2018**.

[100] Betapudi V, Lominadze G, Hsi L, Willard B, Wu M, McCrae KR. Anti-beta2GPI antibodies stimulate endothelial cell microparticle release via a nonmuscle myosin II motor protein-dependent pathway. *Blood*. **2013**; 122 (23); 3808-17.

[101] Wu M, Barnard J, Kundu S, McCrae KR. A novel pathway of cellular activation mediated by antiphospholipid antibody-induced extracellular vesicles. *J Thromb Haemost*. **2015**; 13 (10); 1928-40.

[102] Pericleous C, Clarke LA, Brogan PA, Latchman DS, Isenberg DA, Ioannou Y, et al. Endothelial microparticle release is stimulated in vitro by purified IgG from patients with the antiphospholipid syndrome. *Thromb Haemost*. **2013**; 109 (1); 72-8.

[103] Tong M, Johansson C, Xiao F, Stone PR, James JL, Chen Q, et al. Antiphospholipid antibodies increase the levels of mitochondrial DNA in placental extracellular vesicles: Alarmin-g for preeclampsia. *Sci Rep*. **2017**; 7 (1); 16556.